# Advances
# in
# Human–Computer
# Interaction

# Volume 5

# Advances
## in
# Human–Computer Interaction

## Volume 5

Edited by

## Jakob Nielsen

SunSoft

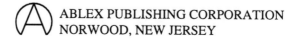

ABLEX PUBLISHING CORPORATION
NORWOOD, NEW JERSEY

ISBN: 1-56750-199-0 (cloth)/1-56750-196-6 (paper)

ISSN: 0748-8602

Ablex Publishing Corporation
355 Chestnut Street
Norwood, New Jersey 07648

# Contents

# Preface

Gary Perlman at the Ohio State University maintains a very useful online bibliography of most of the books and papers published in the human–computer interaction (HCI) field. HCI is usually recognized as having started as a separate field in 1982 with the first major conference, and Perlman's bibliography contains 215 entries for that year. For 1993 (the latest year for which full statistics are available), the HCI bibliography contains 2043 entries. In other words, the literature in the HCI field has grown by a factor of 10 in just 11 years.

One may view the explosive growth of the HCI literature as a sign of the health of the field. Indeed, there are ever more conferences, journals, and books published, and the comp.human-factors newsgroup on the Internet adds about 40% in new subscribers every year. At the same time, the information explosion means that practitioners in the HCI field have become unable to keep up with the literature. Even researchers who need to read other research papers as a part of their job cannot keep up with the literature outside of their own narrow field of expertise. Even in 1982 I doubt that anybody read all 215 HCI papers, but it would at least have been physically possible to do so and maintain a renaissance-like awareness of the complete field.

As editor of the *Advances in HCI* I have decided to use this series to try to fight the information overload problem for HCI specialists. We will concentrate on two types of chapters: those that survey an important area within HCI and provide an overview of recent advances, and those that survey interesting specific design or development projects to show how the state of the art in HCI is actually being carried out. A third category will receive less emphasis with one or two chapters in each volume: futuristic, historic, or personal essays by leading HCI specialists who speculate on important trends in the field.

Chapters 1–4 fall into the category of surveys of important subfields of HCI. First, Kellogg and Richards survey the human factors of information on the Internet. The Internet has received so much press coverage in the last two years that I am sure everybody knows what it is. Interestingly, most newspaper articles about the Internet include some statement lamenting the "cryptic commands and codes" necessary to use it. A first, naive, analysis might lead one to believe that graphical user interfaces would solve this problem by hiding the command names from the user. But as most HCI specialists know, GUIs can have usability problems too, especially when they are retrofitted onto a system with problematic semantics. Thus, would-be developers of the "information superhighway" would be well advised to read (and heed) chapter 1.

Chapters 2 and 3 survey current hot user interface technologies. Bederson and Druin discuss computer-augmented environments, which are a somewhat different take on enhanced interactions than virtual reality with its cumbersome helmets. In chapter 3, Isbister and Layton survey the use of agents as another way of enriching the user's experience compared with the tedious sequence of commands needed in traditional systems. I particularly want to highlight the subtitle of Bederson and Druin's chapter, "New Places to Learn, Work, and Play." All three applications of computers are important for HCI and all can benefit from increased attention to usability. Indeed, because the majority of computer sales now go into the home, home computing, entertainment interfaces, and children's interfaces will probably dominate in the future. Most HCI work (including, admittedly, much of this book) still focuses on use of computers in the workplace, but designing good interfaces for shooting aliens in videogames will soon become a bigger business than designing good interfaces for shooting Soviets for the Defense Department. Even people who prefer nonviolent games probably agree that this is a positive development.

The most classic trap in user interface design is to assume that the user is like yourself. Because most developers (at least so far) have tended to be young, mathematically inclined hotshots, there are definite dangers to this assumption. To increase the sales of computers to a broader audience than other hackers, we have to design with the customers' characteristics in mind. One important market consists of disabled users and Bergman and Johnson discuss ways of making computers more accessible to these customers in chapter 4.

Chapters 5–7 belong to the second category of project reports. In chapter 5, Hunter describes a project at Hewlett Packard to design a pen-based interface to a groupware system, thus combining two important subfields of HCI. The fact that interesting systems like the one described by Hunter invariably draws from multiple fields is an important lesson that should cause some reflection in this age of specialization. Multidisciplinarity is a traditional concern of human factors people, and both Graham and Dye and Turner, Lee, and Atwood comment on the collaboration between usability specialists and developers in their chapters. Graham and Dye describe ways of using information about user needs in the design of Microsoft applications that typically reach a very large market of diverse users and Turner, Lee, and Atwood describe usability work on a much more narrow (but economically important) system for NYNEX telephone company operators. It is instructive to compare these two chapters and their lessons with respect to usability needs for the two main types of software development: sale on the open market and in-house use. A general observation across all three project report chapters is that results, methods, and concepts from HCI research influenced the development of real systems. One often hears HCI researchers complain that system development disregards the research in the field. Of course, it is true that no HCI practitioner can read even a small fraction of the 2043 papers and books published in a year, and it is also true that lack of infrastructure and other practical constraints keep many new technologies from being used in current products. However, the better and more relevant parts of HCI research definitely do not go to waste.

The futuristic essay for this volume of the *Advances* was written by John Thomas, who is an Executive Director at NYNEX and one of the pioneers in the HCI field. One of his early papers was "A Psychological Study of Query by Example" presented with John Gould at the National Computer Conference in May 1975. Now, 20 years later, we *still* don't have really good and usable query interfaces (though we have some decent ones), and one of my current projects is actually the design of the search interface for Sun's online documentation. The search-and-query example shows the value of taking a long-term view of the HCI field and considering where it is going and what major trends have happened and will happen. In chapter 8, Thomas speculates on usability engineering in the year 2020, and I encourage you to consider how his predicted changes may impact you and your work and whether there are some such changes you want to help make come true.

It is in the nature of things that people prefer talking about their successes and forgetting their failures. Even so, many valuable lessons can be learned from considering why things go wrong. In user testing, we often lean more when the user makes an error than when he or she merrily breezes along in the interface. Similarly for projects, it would be good if people would reveal more about those things that did not work. Since it is almost impossible to get people to write about failed projects, I have included a chapter about one of my own mistakes: the independent iterative design project Bergman and I describe in chapter 9. I encourage you to contact me (email jakob.nielsen@sun.com) if you have a failed user-interface project you are willing to analyze frankly in a chapter for a future volume of the *Advances in HCI*.

*Jakob Nielsen*
Mountain View, CA, January 1995

# Chapter 1
# The Human Factors of Information on the Internet*

Wendy A. Kellogg
John T. Richards

*IBM T. J. Watson Research Center*

## INTRODUCTION

"The Internet address you have accessed is currently maintained by Adam Curry. Adam Curry's right to maintain this address is the subject of a pending lawsuit. This address no longer contains Adam Curry's Internet service. The new address for Adam Curry's site is metaverse.com." ["Oh, OK."... user clicks on metaverse.com]

"Unable to access metaverse.com. Its address can't be found."

"Mail returned—host unknown"

"No user—server closing connection"

"This site has moved to somewhereElse.net. Sorry for any inconvenience."

"Nothing returned from your search for 'human factors'."

"Too many connections—try again later."

"You have 377 new items in your mailbox since your last logon [yesterday]. Good luck, pal." (message from a sympathetic mail reader).

*We thank Jakob Nielsen, John Prager, John Thomas, and three anonymous reviewers for helpful comments on earlier versions of this chapter.

If you've been out and about on the Internet (the "net") lately, then the chances are fair that messages such as these are familiar. Perhaps you have already been tempted to conclude that this sort of information constitutes the majority of information available on the net. If messages such as these sound really familiar, you may have been considering giving up the Internet until such time as it can demonstrate better behavior. In the "good old days" of just a few years ago, users could choose to forego technologies that frustrated them—there were viable alternatives, such as going to a library or sending first-class mail. However, avoiding frustration by avoiding the Internet itself is a choice becoming increasingly difficult to live with. There is simply too much going on there; information in electronic form is available to users almost as soon as it is released, and the advantages conveyed on the net-savvy may soon be indispensable—from coordinating picking up the kids, to transacting personal business, to telecommuting to work.

The Internet is a network of computer networks that is growing at a phenomenal rate. The size of the Internet can only be estimated, based on a variety of measures; recent estimates are of over 3.8 million host computers (as of November, 1994), with millions of users (Calcari, 1994; Goodman, Press, Ruth, & Rutkowski, 1994; Styx, 1993). The "information superhighway" is a high priority of the current administration in the United States (Clinton & Gore, 1992), and is becoming a central focus of business (Cronin, 1994) and education (Murray, 1993; Soloway, 1993). It has been claimed that businesses not utilizing the Internet by the year 2000 will have little chance of staying competitive through the next decades (Cronin, 1994).

In its relatively short history, the Internet has grown not just in size, but in complexity and purpose. Beginning with the ARPAnet backbone created in the early 1970s to allow supercomputer resources to be shared by university scientists, today's Internet is a heterogeneous environment of nodes, users, facilities, and services. Enhancements necessary for the privatization of the Internet are well under way. A global electronic society has formed and debate centers on establishing the "laws" of this new land, not whether it should be settled at all. Soon Internet access will be part of all common computer operating systems, throwing open the floodgates of cyberspace to millions of new users. As more and more people and services come online, the human factors of the Internet will become increasingly critical. The current usability of the Internet leaves much to be desired, despite the fact that software and tools with which to explore the net are evolving rapidly. Moreover, there are persistent human factors issues that will need to be addressed no matter how advanced the underlying technology and tools.

Our review of the human factors of the Internet is strongly influenced by the perspective from which we view it—from within a research institution in the United States, as users of the Internet for over two decades, and as designers and developers of graphical user interface Internet software for teachers and kids over the last year. We are thus given to understand many weaknesses in human-net interaction as limitations of the current Internet protocols as much as failures of the designs of current tools which, after all, are constrained to follow the protocols. Our backgrounds as cognitive psychologists and human-computer interaction researchers lead us to try to pick out for analysis the significant and perennial issues for people interacting with information on the Internet.

There are two fundamental aspects of interacting with the Internet: It brings users in contact with a world of information, and with a world of other people. Each of these aspects raises challenges for a user who becomes a denizen of the net. In addition, these factors, by their very existence, and by what is made possible given their existence, raise significant issues for organizations of people and their behavior—that is, for society. This is not surprising; like every major, pervasive technology before it, the Internet is making old practices and socio-economic arrangements impractical even as it is offering a plethora of new opportunities and practices. The Internet is already changing the way we communicate in business, what it means to publish, how we talk to our government, and how and what it means to educate our children.

Our goal is thus to identify and discuss issues that will have to be addressed by any generation of Internet technology. In this chapter, we focus on human factors issues in the arena of people using the Internet as an information source. There are comparable human factors issues in the arena of people using the Internet as a medium for interacting with other people, but unfortunately we are unable to adequately address them here. For the purposes of this chapter, we presume only what we perceive to be the fundamental facts of the Internet: vast amounts of information organized and made available by distributed, situated people and organizations, the ability to interact with this information via some form of technological or human interface, the ability to interact with other people, and the progressively fast migration of business and commercial concerns onto the net.

Given the rapid pace of change of the Internet, we have chosen not to organize the chapter in terms of specific tools/activities that are available now (e.g., Gopher, Mosaic, FTP, Internet Relay Chat, Telnet, Archie). We have also chosen not to critique the human factors of client applications in terms of specific interactions (e.g., selecting or navigating). Rather, we have chosen to characterize the issues in terms of the kinds of opportunities a resource like the Internet affords, and the kinds of tasks with which people will inevitably be faced as they assimilate and attempt to make use of it as an information resource. Our analysis is meant to capture and describe at a relatively high level the central scenarios of information use on the Internet. Of course, with any particular software tool for interacting with the net, we would expect these issues to surface in many contextualized and specific ways. We are unable to provide more than a few illustrative examples at this level of analysis, but we hope that the more general analysis we present may be useful as a framework to others interested in carrying out more detailed examinations of particular tools.

Some have claimed that although the Internet used to be user-hostile (i.e., in the "bad old days" of Unix commands), now, with the advent of graphical user interface tools, the riches of the Internet have been laid bare for the common user. "The availability of these new tools signals the end of the hard-to-use Internet era, and the beginning of a new era that makes access for the masses a reality" (Kantor & Neubarth, 1994). It is true that pointing and clicking, and the aesthetic appeal of graphical user interfaces, makes navigating the Internet more attractive, but, in our experience, proclamations of the end of frustration and usability problems on the Internet are premature, to say the least. Very likely, somewhat less knowledge is now required of users to get connected to the net and begin interacting. Trading off with this newfound ease of use, however, is an explosion of the amount of information and the number of people on the net. The explosion in information means that it can be easier than ever for users to get sidetracked, and harder than ever to find what they are looking for; the explosion in people often means that "traffic jams" on the information super-highway can prevent access to the tools and facilities that users need to accomplish something useful.

## GETTING CONNECTED TO THE INTERNET

### What's Wrong Now

It's easy to summarize what is wrong with getting connected to the Internet: It's too hard. Consider the following scenario:

A user decides to get access to the Internet from his home computer. Because he's an experienced computer user, he knows he'll need a modem, so he gets one and installs it, along with a phone line. Next, he determines what Internet service providers are in his area, and compares the types of connectivity (SL/IP, PPP, dial-up, dedicated line, limited or full access), and price structures offered by each.

Once a service provider is chosen, the user configures his system to work with the service provider . He edits his TCP/IP installation program, carefully entering the data given to him by the service provider (example: your Domain Name Server will be 123.12.3.12). Although the TCP/IP installation program doesn't have a place to specify a "Domain Name Server," he guesses that the "Domain" field is the right place to put this information. The user is grateful that he doesn't have to understand what a domain name server is.

After finishing the installation process, and with great anticipation, the user initiates contact. Nothing. But the user is very motivated; he goes willingly into the process of debugging the connection. He calls a technical friend, who makes numerous suggestions: "Maybe your TCP/IP stack is incompatible with the service provider's software?" "Perhaps your e-mail client inadvertently violates the POP-server protocol?" "Maybe the service provider's DNS was just down when you tested your connection (could you 'ping' it by its IP address?)"

From an end-user's point of view, connecting to the Internet involves learning a great deal of information that is of little intrinsic interest, but which is, at the same time, absolutely critical for making intelligent choices about what kind of connection is needed. Some of the information needed to make good choices (e.g., among available pricing structures) depends on usage patterns that will only emerge and stabilize after the user is on the net, when basic decisions about connectivity have long since been made. Users have little basis on which to decide what quality of connection is needed. Many experienced users rely on a familiar heuristic: Get the best connection you can afford, because sooner or later, you'll wish you had more.

Some users are lucky enough to have access to the Internet from within an organization that has support staff to worry about making the connection work. Users in this situation usually just issue commands to start client programs, and they're on. Usage can get complicated for these users, though, if the organization is behind a "fire wall" (a protective barrier that limits access from outside the organization). Users might find that once they locate something valuable, their system doesn't allow them to transfer files to their local machine. These users are also limited to whatever software clients their systems operators choose or can be convinced to install, which can mean delays in gaining access to useful tools or the latest releases of software. However, they are spared the complexities of getting a working connection to the Internet up and operational, which at the moment is worth a lot.

A variation on this theme is for standalone users to connect to the Internet indirectly through commercial online services. This approach has the advantages of hiding the complexity of getting a working connection to the Internet and of presenting Internet information in a familiar way, but full access to the Internet has been slow in coming, and, in some cases, is still far from complete. Internet services that are provided (e.g., e-mail) are sometimes limited (e.g., no provision for attaching files), decreasing the usefulness of the service. In addition, newer Internet facilities (e.g., WorldWideWeb browsers) may become available, but will not be very usable until the bandwidth of these connections is improved. Users who want to take advantage of higher speed connections to the Internet currently have to do it themselves.

For now, most users may find that connecting directly to an Internet service provider requires too much effort to become an expert on the intricacies of the Internet. Only those willing to acquire technical knowledge of TCP/IP, possible kinds of Internet connections (e.g., dial-in, SL/IP, PPP), IP addresses, sub-net masks, and other esoteric knowledge can be assured that they will be able to make the decisions and do the installation required to make a viable connection with the service provider.

## How it Might be Better in the Future

The process of getting connected is becoming more manageable, and the coming trend of shipping operating systems with TCP/IP installed will help. Packages of "turnkey" client applications are beginning to be offered to users (see Figure 1), and many will be preconfigured to work with certain service providers. Configuring the software to work with other service providers will involve no more than filling out a dialog box with routine information that users can get from their service provider. However, right now this solution does not apply to all users—for example, those who are attempting to use "freeware" or "shareware" clients (i.e., publicly available software where usage is permitted for free, or for a nominal fee). However, even for freeware and shareware, the knowledge burden for do-it-yourself users can be expected to lessen in the near-term future. Authors of shareware programs may create versions already customized to work with various service providers. Then, a user who expected to get her connection through a particular provider would be able to locate a client known to work with that provider.

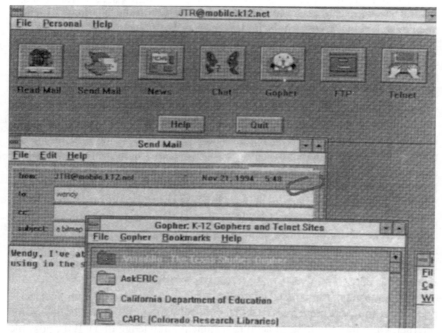

Figure 1. IBM's School Internet software developed by the authors and colleagues at the T.J. Watson Research Center. An example of an integrated "turnkey" package of Internet client applications for end-users. (The School Internet team includes Vicki Hanson, Wendy Kellogg, Petar Makara, Peter Malkin, John Richards, Whitney Rugg, and Cal Swart.)

Nevertheless, we also expect a significant role to be played by commercial application suites, appealing to users both because of their promise of keeping them from having to learn the technical details of connecting to the net, and also for the ease of use and services that will most likely be associated with such offerings. Because the Internet is evolving rapidly, communication protocols and, therefore, software based on those protocols, will also change frequently. For many users we expect the advantages of commercial support services such as automatic software upgrades via the net to outweigh less expensive alternatives that will require finding and reinstalling software every several months in order to stay current.

## DEALING WITH INFORMATION ON THE INTERNET

Historically, the raison d'etre of the Internet according to its developers has been to share data and computing resources. As such, much of its value has always been thought to lie in simply making information (or computing resources) available. This claim has been countered by actual experience in the deployment of telecommunications networks around the world, which suggests that the ability to communicate with other people is the strongest motivation for gaining access to the Internet (Rheingold, 1993). Still, there is no doubt that users do seek and use information of value to them. Here we describe some of the fundamental human factors issues in finding and utilizing information from the world-wide net.

### Discovering Information

Descriptions of what it will be like for users to find and access information via the Internet often seem overly optimistic, and to contradict years of empirical research on information retrieval (Dumais, 1988). Consider, for example, the portrayal of "a researcher with a query, seated at her networked workstation" in the following scenario: "From a friendly user interface, she connects to [a] library's welcome menu offering icons and introductory text. When prompted, she enters her question, and then waits for an expert system to process the request. Almost instantly, a selection of materials on her topic appears" (Clement, 1994, p. 62).

From an HCI perspective, what is not mentioned in this description is as important as what is said: for example, the recognizability of the icons, the language and syntax in which the query must be expressed, the ability of the expert system to deal with synonymy or context, the length of the wait for the expert system to process the search (assuming the expert system can be accessed without delay), and the selection of materials *not* on her topic that is also returned, "almost instantly." Unfortunately, the realities of discovering information for users are often closer to what is not mentioned in the earlier description.

In the earliest days of the Internet, users could only find things by navigating to a particular server and then looking around. There was absolutely no way to ask the question of whether information on a particular topic was available, and if so, where. All a user could possibly do was collect the names of promising servers and visit them periodically to see if anything new of interest had appeared. The importance of being able to discover useful information on the Internet and the impracticality of searching world-wide file archives by trial and error is a long-recognized problem. Deutsch, one of the co-inventors of a search facility called Archie, has referred to this as the "resource discovery problem" (cited in Dern, 1994, p. 331). In 1989, Deutsch and colleagues developed a program called Archie, reflecting their frustration with the "go-there-periodically-and-see" method of finding things, and the realization that an automated tool for collecting the contents of publicly accessible FTP servers was possible based on existing mechanisms (Deutsch, 1994). Archie allowed users to use simple keyword searches to locate FTP-able files by name matches. For the first time, it became possible to search for information of interest without having to go to the server on which it was stored.

With the advent of Gopher in 1991, opportunistic browsing of Internet resources became possible. By presenting a "menu" of choices representing directories of various servers, Gopher opened up access to a wide range of servers without the user having to know their names in order to go there. However, users did still have to go to a server to see what was being offered; Gopher did not provide a means to search, other than by looking around. Search services were not long in becoming available for Gopher, however. In 1992, a search facility similar to Archie, called Veronica, became available for Gopherspace. Veronica was soon followed by Jughead—a variant that limits its searches to local rather than Internet-wide Gopherspace. More recently, the WorldWideWeb has spawned a large number of search mechanisms based on programs that periodically "crawl" through the web collecting the names and content of WWW servers (e.g., the Web Crawler, WorldWideWeb Worm).

Although there has been a steady increase in the number of tools available to help users find information, existing tools suffer from well-studied limitations, as discussed later. For many users, we suspect that less formal methods, such as browsing or asking other people, may be the most frequently used discovery techniques. The core methods for discovering information are opportunistic exploration, searching for information, and asking others.

***Opportunistic exploration.*** In some ways, the lowest effort strategy for discovery is to simply look around. Prospecting for information on the Internet opportunistically, however, is largely unprofitable; in our experience, the typical "return on investment" (i.e., amount of value gained for time spent searching) is low.[1] Even when something of value can be located, users often find it takes too much time to achieve the payoff of having the found reference in hand.

Gopher, and WorldWideWeb browsers (Figure 2), are examples par excellence of methods available to users for discovering information on the net opportunistically. Both allow users to pursue interesting-looking resources without having to know where they are located. There are two ways to do this—by browsing around from one "page" of information to another in either information space, or by searching for "pages" on particular topics via search mechanisms that can be accessed from within the space. Browsing is aided by the common phenomenon of collections of related pointers (selectable links pointing to Uniform Resource Locators or URL's in WWW, menu items pointing to Gopher servers in Gopher). Such collections are usually put together by other Internet users to serve particular groups of users or interests. For example, a Gopher collection known as "Gopher Jewels" maintains a group of pointers to exciting uses of the Internet in K-12 education, and other resources (e.g., curriculum ideas) of interest to K–12 educators. Once a collection like "Gopher Jewels" is assembled, it can be replicated as a menu item on many Gopher "pages" throughout Gopherspace. Collections greatly improve the chances of finding information opportunistically. A downside of collections versus searching for information, however, is that users will only be able to find newly available material if the collection owner continually searches and updates the collection—on the other hand, searches are accompanied by their own challenges.

*Searching for information.* Gopher, FTP, and WWW have search mechanisms available. In Gopherspace, for example, Veronica allows users to search for Gopher items matching a keyword or phrase. Internet search mechanisms are limited by access; users who want to perform a search must first find an available Veronica server to run it. Typically this involves repeated requests to a remote server (or servers), hence the familiar "Too many connections—try again later" message. Currently, access seems to be more of a problem for Veronica servers than most WWW search facilities, but the access problem for all search mechanisms can be expected to increase as the number of new Internet users outpaces the availability of new search sites. Another problem has resulted from attempts to limit the domain of searches or the maximum number of items returnable—a kind of user-controlled "triage" intended to allow Veronica servers to serve more users in less time. Before entering their search term, for example, Veronica users now find themselves faced with a long list of specialized search choices that vary on what will be searched (e.g., directory or file titles, "white pages," etc.), and the maximum number of items that will be returned (200, 400, 1,000, etc.). Unfortunately, the difference among these

---

[1]This is not to say that it isn't easy enough to find interesting things to get sidetracked by, only that attempting to address specific questions or more general work goals is often harder than one might expect.

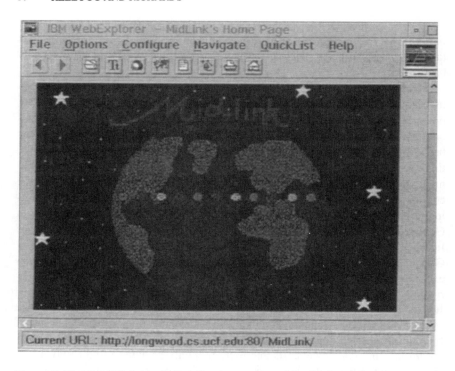

Figure 2. The IBM Web Explorer, showing an example of information a user might discover opportunistically. Midlink Magazine is an electronic magazine created by middle-school kids (ages 10–15) for other kids.

search choices and the reasons for choosing one over another in particular circumstances are left for users to discern on their own.

Keyword search as a method for efficiently finding information of interest suffers from well-known problems and limitations (see Dumais, 1988, for a review). Studies of information retrieval show that, typically, only about 50% of what is returned from keyword searches is wanted (the "precision" rate), and at the same time, only about half of what exists in the database that is wanted is found (the "recall" rate; Dumais, 1988). As Dumais pointed out, low precision may not be a problem when only a handful of items are returned, but when hundreds or thousands of items are retrieved, the large number of irrelevant items is problematic. These estimates of recall and precision, based on controlled studies where the characteristics of database items are known and a range of query sophistication has been employed, can only be optimistic for Internet search engines where the syntax of queries has been kept deliberately simple (the default being to search the database for user-specified keywords that the search engine "AND"s together), and the "database" is vast and irregular. Although more complex queries are usually possible, learning how to make them requires locating documentation on the capabilities of the particular search mechanism being used—which also assumes that the mechanism is identified to users, as in the following warning for searches of the World-Wide Web Worm (WWWW):

> WWWW currently uses the UNIX egrep command to perform database searches. If you are not familiar with egrep, it is advisible to restrict to single keyword searches. Also, because the database being searched is huge, expect to wait a while.

Note that this message, while helpfully providing the information of what search engine is being used, does not tell users where they can access documentation on the egrep command online if they want to learn more. It is interesting to contrast this message with a similar message for Web Crawler users (Figure 3), which advises: "This database is indexed by content. That means that the contents of documents are indexed, not just their titles and URLs. Type as many retrieval keywords as possible; it will help to uniquely identify what you are looking for."

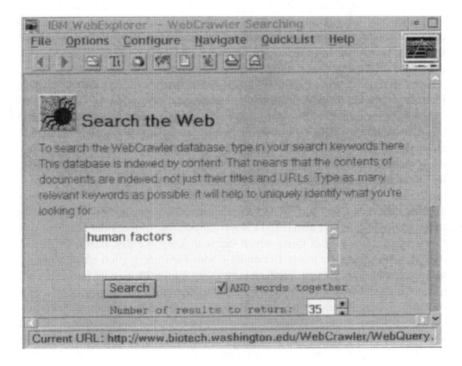

Figure 3. The Web Crawler, one of the search facilities available on the World-WideWeb http://www.biotech.washington.edu/WebCrawler/WebQuery.html).

The "Worm" and the "Web Crawler" search engines appear to be based on different information retrieval schemes, therefore the most effective search strategies are different. Veronica and Archie search engines may be different still. For users, the diversity and incompatibility of Internet search engines signals a low likelihood of good results unless significant effort is invested in learning a particular retrieval system well. Another consequence of the inability for most users to search (relatively) optimally, coupled with the sheer magnitude of material available on the net, is that retrieval will be more haphazard than it ought to be—instead of finding the most relevant items available, users will often "satisfice" on the first items they encounter of any relevance. It also means that FAQ (frequently-asked questions) files and messages such as was just presented offering search tips and advice will continue to be the norm. As Norman (1992) has pointed out, "signage" posted alongside technology is often a clue that the design is less than optimal from a human point of view.

Information retrieval is a well-studied aspect of human–computer interaction, and much more is known than is currently implemented in tools for Internet searches. Exploring the use of some of these more effective approaches in future tools could have great benefits for users. There is no shortage of ideas for improving the search and retrieval capabilities of browsers: Developers might begin by looking into techniques such as unlimited aliasing, adaptive indexing, latent semantic indexing, relevance feedback, "fisheye" views, rich indexing, information lens filters, dynamic queries, aggregation mechanisms, and the use of intelligent agents (Ahlberg, Williamson, & Shneiderman, 1992; Dumais, 1988; Goldstein & Roth, 1994; Landauer, 1991).

*Asking others.* Finally, information of interest can often be discovered through communication with other people. Newsgroups and e-mail-based discussion lists such as LISTSERVs provide access to a group of people with common interests. Members of a discussion often post items of interest to the entire list, facilitating opportunistic discovery by other list members. Many lists maintain "FAQ" (frequently asked questions) files pertaining to their focus of interest, and these can be a valuable source of information, especially for newcomers to the discussion. FAQs also help to lower the number of redundant queries that the list entertains across time, which helps to keep the discussion fresh and moving along. Finally, interest-oriented discussions are a good place to ask a knowledgeable group of people for pointers regarding the kind of information a user is interested in finding. Discovering information by asking people is a mechanism that already works well (Bannon, 1986), and other people will continue to be one of the best resources of the Internet for finding things of real value. It might be interesting to explore the use of ephemeral interest groups (Brothers et al.,1992) among Internet users to facilitate the formation of temporary "information hunting" groups for people looking for the same sort of information. Tutorials and courses via the Internet are already being offered to help new users learn to use the Internet effectively. In the future, we expect to see more of such activity and efforts to "coach" other users in getting the most out of the Internet. Even when users succeed in finding (seemingly) useful information, however, their job is not done: They will have to face significant challenges in order to bring the information into their local computing environment.

## Fetching Information

When there is an almost limitless amount of material that could be transferred to one's own computer, it quickly becomes clear that a fundamental issue is deciding when it is worth it to download something. It does not take long to stress the storage capacity of one's local computing environment, or, perhaps even sooner, one's ability to deal effectively with information gathered indiscriminately. On the other hand, using the Internet as a storage medium has a major disadvantage, because users cannot control whether something they want to refer to in the future stays available. There is thus a constant tension between wanting to preserve valuable information in a place where its persistence can be controlled, and leaving it out there on the net, keeping a pointer, and hoping that it will still be there someday when it is needed.

Users of information on the Internet will need to address several questions before and as they come in contact with information resources. The core tasks of fetching information are: understanding and formulating information-retrieval "values," evaluating particular pieces of information, deciding on a plan of action, and recovering from information retrieval failures. The latter three tasks can be viewed as roughly corresponding to some of the stages of user action described in Norman (1986).

*Understanding and formulating information-retrieval "values."* Users need to come to understand the kind of information available on the Internet, its likely longevity, and so on, and develop a set of practices for dealing with it. These practices should map to the possible actions that can be taken with respect to fetching information. Often, it is experiences of the ease of losing track of once-found information that help to shape a user's "policy" about handling information.

However, software tools will also shape a user's "policy" for handling information; after all, they determine which actions are possible to take towards particular pieces of information. A user's mental model of goals it is possible to have vis-a-vis Internet information will arise, at least in part, from the functionality of the software (Carroll & Kellogg, 1989). For example, a Gopher client that supports viewing, bookmarking, and transferring files suggests these as rudimentary evaluative categories, ordered by their increasing ability to preserve information under the user's control.

As users grow more experienced with Internet information, however, their evaluative space will expand; usually far beyond the support capabilities of current software. For example, the user may begin to feel the need to store bookmarks representing collections of pointers separately from those representing "pages" of information on particular topics. Users might want to store related information (e.g., issues, or selected parts of issues of an online periodical) in a single place, under a single title. As this collection—the user's personalized view of the Internet—grows, so does the need for a structured storage system and more sophisticated ways of browsing and retrieving stored pointers or Internet data.

*Evaluating particular pieces of information.* A user's "retrieval policy" is an important guide, but equally important are the means available to evaluate promising information. Every time a user encounters what seems to be an intriguing piece of information—say, a link in WWW, or a file in Gopher or FTP—that information must be evaluated. The user must, in effect, ask, "What should I do (if anything) with this piece of information?" Of course, users may be in different states when they encounter information: On some occasions something just draws their attention and starts the evaluation process (psychologists might characterize this as a "bottom-up" triggering of evaluation); at other times, users may be looking for something that can fulfill a particular purpose and will want to evaluate the goodness of fit of the information with their purpose (a "top-down" initiation of evaluation). Whatever the nature of the evaluative process, however, it will be the outcome of it that leads to users establishing the goal (Norman, 1986) to fetch the information.

The evaluative process itself, however, is not always well-supported. A sensible approach to encountering potentially useful information might be to find out enough about it to know whether it makes sense to commit to it further—for example, to visit it or transfer it. Often, however, the information available to users about what lies at the other end of a WWW link, or the contents of a file, is inadequate to support this kind of decision. In the worst case, there is no information available other than a name or phrase, and users have little choice but to follow the link or fetch the file in order to see whether it is of interest. In other cases, there may be descriptive information close by in the environment. WWW links currently represent the best case, as they are usually displayed in the context of a "page" of information designed to help users decide which links to follow next. Gopher and FTP represent intermediate cases. Directories often contain a README file that describes or indexes the directory's files. Gopher, of course, allows users to display these files; FTP clients often do not. In the design of our School Internet software, we included the ability to display a file in our FTP client specifically to lower the cost to users of getting at descriptive files; often the first thing the user wants to do is to read the README file to see if the directory is likely to contain anything of interest. Allowing users to do this within the FTP client allows them to intermix browsing and fetching activities within a single context. In Gopher, moreover, we allow users to cancel a transfer of information in progress and see the results so far. This gives users a way to quickly verify from a glimpse at a file's content that they are getting what they think they are getting.

However problematic evaluation is from a single user's point of view, from a systems point of view it is worse. When millions of users are forced to follow links or transfer files in order to see whether or not they are interested in the information, there will be an obvious impact on the network's communal bandwidth. How much of an impact is difficult to estimate, but the precise amount may not matter much: At the rate at which the Internet is growing, it may be prudent to eliminate as much superfluous traffic as possible. What is needed are less "all-or-none" ways for users to assess information content; richer ways for them to preview information. One way to do this is to allow users to initiate and quickly cancel a transfer to view a sampling of a site's or file's contents, as we described earlier.[2] Another way is to explicitly code descriptive or summary information with objects that is accessible to clients without having to transfer the entire object.

The display of partial or descriptive information may be a particularly valuable technique when users are seeking information for a particular purpose. It may also be more relevant to applications like FTP, where there is little context for available material, than to applications like WWW browsers that offer material embedded in descriptive content. For example, we were recently seeking images of animals for a special demo. We were looking for close-up images rather than long-range "establishing" shots. Most of the archives we examined employed the conventional practice of supplying an eight-character descriptive name. In the best of cases, this name was interpretable as an animal image (e.g., "FLAMINGO.GIF"); in other cases, it was hard to tell (e.g., "AUSPAR1.GIF"). In no case could we tell until we transferred and displayed the image whether it would meet our criteria. If, on the other hand, our FTP client could have displayed a small icon of the image (ideally with image-size information), it might often have been enough information for us to make better decisions about whether or not an image was interesting enough to warrant a closer look. (Of course, whether such a scheme would be practical given the performance decrement such a convenience would surely entail is another matter, but this becomes less of an issue as bandwidth and processing power increase.) Because image files can be large, the savings in bandwidth from influencing even a small proportion of user decisions could be substantial. Perhaps even more importantly, improving the ability to evaluate the content of files might mean that users would be less likely to waste their time and money waiting and/or paying for files that do not really interest them.

*Deciding on a plan of action.* Once a user has decided it is best to bring a given piece of information into the local environment, there may be a variety of ways to obtain it. Deciding on a particular method of obtaining it corresponds to

---

[2]Note that many web browsers also allow cancellation of a URL load, but typically they do not show users partial results (i.e., the information gathered so far). Thus, although they support users who wish to give up on a transfer because it is taking too long, they fail to support (as they might) the equally plausible goal of previewing information.

Norman's (1986) stage of *forming an intention*. Files discovered on a WWW page, for example, may also be available via e-mail or an FTP site. FTP requests might be set up for later execution; for example, some online services give users the option to schedule downloads to occur at times of low usage and phone rates (e.g., 4:00 a.m.). Many e-mail discussion groups are also available as newsgroups or are posted to Gopher servers, facilitating use of that information by users whose needs can be satisfied by browsing, or who cannot afford to process the floods of mail that sometimes occur in e-mail discussions.

Currently, users need to be alert for information pointing out the existence of alternative ways to get information. If the provider of the information does not announce the alternatives, users have no reliable way of finding out about it. We have experienced Telnet sites whose connections were all occupied that helpfully listed dozens of mirror sites, only to have the server close the connection quite literally before we could grasp the information. Helping users discover other ways to achieve their goals likely could have benefits for Internet traffic as well. For example, if alternative site lists were identifiable objects, and servers had the conceptual equivalent of traffic directors, it would be possible to periodically reorder the lists to keep the lowest traffic sites at the top of the list. In the ultimate, when a user requested a particular information object (say, a file), the server would either send the file itself or delegate the request to a less-busy server. In either case, the user would get the file, but in a manner in which resources could be managed to optimize characteristics such as response time, bandwidth consumption, and so on.

***Recovering from information-retrieval failures***. When an attempt to retrieve information fails, users need to assess what has happened and what recovery options might be available. This task represents one aspect of what Norman (1986) called *evaluating the system state with respect to the goals and intentions*. For Internet retrieval, the problem generally tends to be more one of frustration for the user than an inability to assess the source of the failure or the options for repairing it. Indeed, perhaps it is because the cause is transparent that having to pursue the remedy provokes frustration (Carroll & Kellogg, 1989).

Two small examples may suffice. Usually when a request to load a URL or to get a file fails during transmission, the user must start over and re-fetch the information. Having to repeat one's request can be mildly frustrating, but vexation increases directly with the total amount of information to be fetched and the proportion that has already been transferred at the time of failure. In newer environments, such as WWW browsers, users are able to view the hypertext information as soon as its transmission is complete, allowing images to continue coming in on a separate stream. This may lead to cases in which the links on the page are usable even when transmission fails before the images are complete. Older applications, however, tend to be "all-or-none": If transmission fails before the transfer is complete, it may be impractical or impossible to display what has been successfully fetched, or to transfer only the remaining portion of a file.

A second example is when there is insufficient disk space to receive a file, but the user is not informed of this until late in the information-transfer process. In this case, the cause is often explicitly communicated in an error message (e.g., "Unable to complete transfer; out of disk space"). An obvious human factors recommendation is to warn the user before the transfer commences that there is insufficient space at the requested target location, but, depending on the protocol being used and the client software, it may or may not be possible to do so. File transfer within the Gopher protocol does not support getting any information about file size in advance of the transfer, whereas the FTP protocol does support it, but relies on the client taking advantage of the information in order to provide a warning.

These examples can be seen as raising implicit questions about the scope of responsibilities of Internet protocols, servers, and clients. The first case points to the ability of these to support fetch requests, which works well as long as the transmission is successful. When it is not, neither clients, servers, nor protocols have the responsibility of ensuring successful completion of the request from the client's (or user's) point of view; rather, it is up to the user to resubmit the request. Similarly, in the second case, Internet components are not designed to interact cooperatively with the state of the user's local environment in fulfilling a user's request. We discuss the implications for Internet protocols of improving the human factors of the Internet later.

**Using Information**

Currently, one of the most vexing aspects of the Internet is trying to actually use information once it has been obtained. We refer not to the user's ability to profit from the content of the information per se, but to see, hear, or run it in the first place. Too often, Internet client applications that allow discovery and recovery of information do not include the ancillary software necessary for users to actually experience the fruits of their labor. We call such software "instrumental" because, from the user's point of view, it is of interest and value only due to the fact that it is needed to display or run something that is inherently of interest. Core tasks for using Internet information include dealing with instrumental software and evaluating usage failures.

*Dealing with instrumental software.* Users need to learn when instrumental software is needed to use files gathered from the Internet. Image files require the right viewer; executable and other binary files may need to be decompressed. When information is compressed, as it often is for more efficient transmission, the user will need the appropriate tools for unpacking and/or viewing transferred files. Knowing which tools are necessary means learning the file-labeling conventions (e.g., "aFile.zip" needs to be decompressed with the "unzip" program), and/or reading information posted at the same location as the files (e.g., README files). Sometimes, servers will decompress a file on the fly for users, if they ask for it by a name that does not include the compression extension (e.g., requesting "aFile.txt" instead of "aFile.txt.zip"). But in order to take advantage of this, users must discover that the service is available, and their client software must support requesting a file by name (e.g., instead of only by pointing and clicking, because usually the uncompressed file name is not listed in the directory).

Users must do more than understand what instrumental software they lack; they must acquire and install it before they can use the files that require it. Most convenient for users, when their environment does not contain a full complement of instrumental software, are sites where the relevant software is made available along with the files offered that require it for use. Some sites have a README that directs users to archives for instrumental software. Others just post files, and users are forced to discover by other means why the files they transfer "don't work."

Tools have to be installed so that they function correctly and can be accessed conveniently. Ideally, they will be invoked automatically where appropriate, as a function of file-type or file-header information. For many types of familiar desktop applications, recruiting the appropriate application when the user opens the file from the desktop is automated by the operating system. For files downloaded from the Internet, though, users typically must launch the appropriate program themselves and then open the file from within the application.

A partial solution for the future is to build instrumental software into the user's local environment—whether in client applications or (probably better) into the operating system. Viewers and compression/decompression programs from the user's point of view are just like TCP/IP, or a machine that reads diskettes of any format. Tasks that users perceive as incidental to their use of information should be redesigned out of existence. The goal is: If the user can get it, the computing environment can display it.

*Evaluating usage failures.* When a user's attempt to use a file from the Internet fails, it becomes necessary to determine the cause of the failure—a cognitive task that has been termed *credit assignment* (Minsky, 1961; Schank, Collins, & Hunter, 1986). For example, a user who FTPs a piece of executable software may find that it doesn't run properly, or at all. Discovering the reason can be difficult because there are many possible causes of failure: It may be that the file was downloaded in the wrong "mode" (e.g., ASCII instead of binary), that an inappropriate version of the program was downloaded, or that something in the user's local computing environment conflicts with the operation of the program. If the user discovers a downloading error, he or she may additionally be faced with the problem of reconstructing where the file came from.

Another example occurred to us recently as we spent the better part of an afternoon, on and off, attempting to download and print a postscript file via the WorldWideWeb. We were seduced into the attempt with ease: The file contained a paper that was a perfect reference for this chapter; there was a fine postscript printer available through our facility's main computing systems, and an easy way to transfer the downloaded file from the pc to the mainframe. What could be easier?! When the first attempt to print failed, there followed a series of attempts to modify components of the "download and print" scenario. Our behavior took on familiar dimensions: the growing tide of frustration and then determination to triumph over inexplicable technical difficulties, the creative ferment of the pitiful little hypotheses that seasoned end-users have learned to make (conditioned on a variable ratio schedule that would impress any student of operant conditioning paradigms). "*I know*! Maybe the file extension has to be 'ps' for the printer to recognize it as a postscript file. . . ." Explanations and remedies were imagined and implemented, including two full downloads from the net. The file was not to be printed. The question is, of course. . . why not? At what point did the attempt to retrieve and print the information break down? There was no way to tell. This is a classic example of the problem of credit assignment, one which is made more onerous by a complex, multicomponent system. Unfortunately, even with simpler configurations of technology, this experience is all too common.

There are possible approaches to easing the difficulty of credit assignment, some of which have been in the HCI literature for over a decade. For example, intelligible error messages that suggest remedies are helpful (duBoulay & Matthew, 1984; Lewis & Norman, 1986; Shneiderman, 1980). Programs or other objects that are able to disclose useful information about themselves (Kellogg, Carroll, & Richards, 1992) could assist users in understanding prerequisites (e.g., following the above example, the user could query the postscript printer and find out its prerequisites for printing postscript files, including that the file extension has to be "psbin," or that it doesn't matter, etc.). For Internet users, simple diagnostic tools might be helpful; for example, tools that could tell users whether an object fetched from the net is "well-formed" (i.e., answer questions such as "is this a legitimate postscript file?").

**Remembering Information**

Users have a multidimensional relationship to information available on the net; moreoever, this relationship is dynamic, where the value of information changes over time as the user's interests or goals change. The information itself is also dynamic, both in the sense that the same information can change location (be moved to a different server) or be replicated across different locations, and in the sense that it may be edited or updated over time. These underlying characteristics of users, information, and the evaluative function relating the two give rise to the need to support not only the immediate use of information, but also the *future* use of information. Erickson and Salomon (1991) discussed "practices of information users," emphasizing the need to accommodate in electronic media such activities as skimming, annotating, and organizing information. Here we discuss the following core tasks of remembering information: finding things again, preserving information for future use, and keeping up with information that moves.

*Finding things again.* Users will not always know at the time they encounter information that they will want to find it again. When they do know they will want to refer to something in the future, they are in a position, of course, to take steps to remember the information. At other times, however, they may remember seeing something relevant to their current concerns that they would like to find again. In both cases, technology can provide support

Many clients provide support to users who want to remember how to find something again through facilities to create pointers to the information in the user's local environment. For example, clients may allow users to "mark" information of interest by placing a "bookmark" (Gopher), or by adding a page to a "hotlist" (WWW). These Internet sites can later be reinstated by reactivating the pointer (i.e., opening a bookmarked Gopher site, or a URL from the hotlist). There are limitations to this approach, some of which are discussed in the next section. One limitation with particular relevance for finding things again is the robustness of the pointers, which depends, among other things, on how the function is implemented. Clients intended for use on machines shared by many users may be forced to keep bookmarks or hotlists on less permanent media, such as floppy diskettes. We have also seen clients make the mistake of saving new bookmarks or hotlist elements only on cleanly exiting the program. In both cases, users have a higher risk of losing their pointers. When they are lost, users are forced to start over in creating some or all of their personal-interest-centered map of the Internet. This has the dubious psychological value of providing an opportunity for the user to "rehearse" her favorite sites, but in general, it is probably a better practice to make the user's tags as permanent and hard to lose as possible.

Less common are clients that provide support for finding things again when users have not thought to mark them explicitly ahead of time, a situation that might be described cognitively as ad hoc memory retrieval (cf. Barsalou, 1983). In such cases, users will be forced to recall where they saw the information they remember, or engage in problem solving strategies to identify possible candidate sites at which to look for the information (e.g., thinking about recently or frequently visited sites, or thinking about sites that "ought to" have a pointer to the desired information). An intriguing possibility for supporting ad hoc memory retrieval is for the system to passively record and make visible to users where they have been on the Internet . The IBM School Internet software provides such support on a limited basis by passively recording and making available to users, within the scope of gopher session, all "recently visited" Gopher sites. Another fascinating possibility is to apply the concept described by Hill and

Hollan (1992) as computational "wear":

> Wear is a gradual and unavoidable change due to use. As a source of useful information, wear is particularly appealing since it is a by-product of normal activity and thus essentially free. No extra effort, nor scheduling of additional tasks are required to get its effects. (Hill & Hollan, 1992, p. 7)

One of the examples Hill and Hollan discussed is that of "Menu Wear," whereby the "statistics of previous menu-selections by category of user and by category of context get painted onto the menu items themselves." In the case of a Gopher client, this would allow users to see, over time, the emergence of well-worn paths through Gopherspace, possibly providing useful clues about where to find information that was not intentionally bookmarked.

*Preserving information for future use.* Users will not always be willing or able to pursue interesting prospects at the time of their discovery. When they cannot, they need the ability to create and manipulate information about Internet sites and pointers that they come across. Over the course of time spent using the Internet, the proportion of information the user interacts with that is "preserved for future use" information (i.e., information that has been previously discovered by the user), as opposed to "just discovered" information, can be expected to grow to account for perhaps the largest proportion of user activity. As such, the task of preserving information for future use is a highly significant and central scenario of Internet use—what has been termed a *core scenario* by Carroll, Kellogg, and Rosson (1991).

Surprisingly, this core scenario receives only meager treatment in current technology. One reason for this state of affairs is that, quite rationally, current tools have tended to optimize the task of navigating to and viewing Internet information, but, not so rationally, simultaneously (and unnecessarily) virtually ignored the quite different goal of assisting users in preserving information for later use. To take one example, most web browsers only weakly support users who know they are interested in a site, but do not wish to go there immediately. It is true, as discussed earlier, that there is usually a "hotlist" whereby the URL can be "bookmarked" for later use. But in order to add the URL to the hotlist, the user usually must *go there first*! This is prima facie evidence of the extreme bias toward navigation and display present in most tools.

Second, the tools that typically are provided to assist users in this core scenario—pointers with uneditable verbal labels—are woefully inadequate with respect to the role they need to fulfill for users. In a brief summary of memory and attention, Miyata and Norman (1986) stated:

> Long-term memory (LTM) consists of organized knowledge units, called schemas, that structure knowledge and also contain the procedural information necessary to control actions. Several aspects of LTM are important, including the difficulty of acquiring new information and the problems and

issues in retrieval of information, once acquired. An important aspect of memory is "reminding," the manner by which one event may cause retrieval of the memory for another. (See Card, Moran, & Newell, 1983, for further discussion of approximate models of memory; see Norman & Bobrow, 1979 and Schank, 1982, for a discussion of reminding and memory retrieval. (p. 266)

Stored pointers such as bookmarks and hotlist elements essentially serve as reminders for users about information they have previously discovered that they thought was interesting or would have future value. The key to their effectiveness in fulfilling this role is how well they serve as a memory retrieval cue, and this, in turn, depends on the quality of the reminder as a partial description of the to-be-remembered materials (Miyata & Norman, 1986). As currently implemented, pointers typically have verbal labels given by the information provider (i.e., not by the user), are uneditable, cannot be annotated, do not conveniently support saving the descriptions that are often found on the Internet, and are presented in flat, unstructured lists. As such, they lack several attributes that would allow them to function as better reminders (Erickson & Salomon, 1991; Miyata & Norman, 1986; Schank, 1982).

WWW users who want to enhance the descriptive information they preserve for a pointer can, of course, do so manually: They can assemble an annotated file (e.g., by copying and pasting the descriptive information, pointing at the link to make the URL visible in the browser, writing it down, and finally typing it into the file), or, at significantly less cost in time and effort, they can simply resort to keeping a (paper and pencil) notebook of interesting sites. The IBM School Internet software improves on this scenario mildly by providing a place where users can annotate site addresses if they choose (i.e., an Address Book into which server addresses can be cut and pasted, and labeled with a user-chosen "nickname" or other brief annotation).

Knowledgeable users who wish to overcome the limitations of pointers that exist in flat, unstructured lists, can, of course, also do so at their discretion. We know of one user who went to the extreme of introducing fictitious URLs into his hotlist (that were, in fact, dashed lines with topic labels), in order to segregate the real URLs into categories of interest to him. Furnas and Zacks (1994) described a structure for representing information called *multitrees* that might be used to organize, represent, and manage access to a personalized view of Internet information. In any case, most users will soon exhaust the capabilities of unstructured lists for representing their personal view of interesting Internet information.

Users do not engage in the activities of viewing, retrieving, or storing information in isolation, nor for their own sake; rather, these activities are part of a larger context of preserving and using information in their work or personal lives. Card, Robertson, and Mackinlay (1991) argued that the real goal should be to optimize the "cost structure" (for users) associated with this larger process of information use. They describe their goal thus:

> If we want to move beyond information retrieval, narrowly conceived, to address the amplification of information-based work processes, we are led to try to develop user interface paradigms oriented toward managing the cost structure of information-based work. This, in turn, leads us to be concerned not just with the retrieval of information from a distant source, but also with the accessing of that information once it is retrieved and in use. (p. 183)

The Card et al. (1991) analysis suggests that until the overall cost structure of preserving and using Internet information is addressed by the tools in the user's computing environment, users will continue to have problems using their tools in a useful and productive way. Nevertheless, it is not hard to envision ways in which tasks could be better supported now. To resume the previous example of preserving descriptive summaries and links to WWW sites, a minimal approach might be to make it more convenient for users to save this information.

Suppose, for example, that a user who encountered a descriptive paragraph with one or more active links could invoke a special "save reference" function that would copy the entire description, whatever its form (e.g., text and image), and preserve the link(s) in a visible and usable form (e.g., include the URLs from the underlying HTML source). Suppose further that the client supplies a specially designed space, separate from the hotlist, for keeping and working with these annotated pointers. The notebook described by Erickson and Salomon (1991) is one conceptualization of such a function. Perhaps, ultimately, there is little distinction between this preserving space and the browser itself, with the "save reference" function effectively copying parts of HTML source files and allowing the user to compose and manipulate these on the fly into custom sets of pages or his or her own use or to share with others as his or her collection of references grows.[3] In other words, imagine Internet tools that address the overall cost structure of information usage for their users, that engage in the same level of sophistication and support for the core scenario of *preserving information for future use*, as they already do for the core scenario of *navigating to and displaying information*.

***Keeping up with information that moves.*** Information on the Internet will always be dynamic: Information within servers will get reorganized as it changes and grows, and information among servers will move or be replicated as people move, organizations change, and so on. Ideally, from the user's point of view, the location of particular pieces of information is irrelevant. Information moves from place to place on the Internet transparently. Practically,

---

[3]We thank Carl Kessler for suggesting the essence of this idea.

this ideal may or may not be achievable with enhancements to the infrastructure.[4] Take, for example, the case of Adam Curry's WWW page moving to a different server. It is easy to imagine that instead of a human-readable message being located at the dead end of the "old" URL, a machine-readable "forwarding message" is found. Instead of the user receiving a message apologizing for the inconvenience and requesting that the URL be updated, the old URL is automatically updated by an (invisible) communication between the server and the user's browser (and is updated everywhere it occurs in the user's environment), and the new URL is simply displayed. The user gets where she was going, and the information gets to move freely.

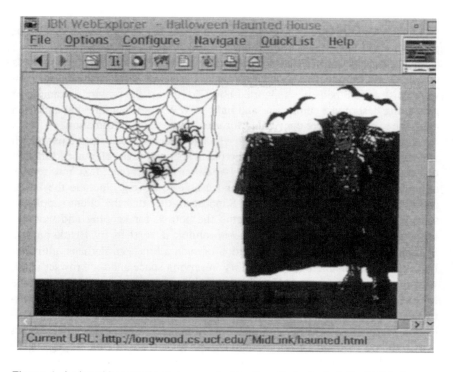

Figure 4. A virtual haunted house, created by the all-student staff of Midlink Magazine at Discovery Middle School in Orlando, Florida. An example of shareable information on the Internet.

---

[4]Currently there is an IETF (Internet Engineering Task Force) working group looking into a possible solution to the problem of location transparency. The working group is investigating the possibility of defining a name space of persistent object names called URNs (Universal Resource Names).

## Sharing Information

Internet users want to share interesting things they find with friends and colleagues around the world (see Figure 4 for an example). We distinguish between the tasks of sharing pointers and structuring information for use by others.

*Sharing pointers and groups of pointers (bookmark lists, "hotlists").* Right now, this is an ad hoc task, making it difficult both to share and use information about Internet discoveries. Typically, users are forced to describe to each other, in detail, their starting place, and the path taken to navigate to a particular reference, or the server's name and path/filename. The URLs (Uniform Resource Locators) of the WWW modestly improve on this scheme, at least allowing an interesting place on the web to be *named* by a single entity (although many times this "name" is indistinguishable from a path specification). At present, however, although users can mail the "name" of a web page (i.e., the URL), they cannot mail the link itself (or a set of links). People receiving a hotlist or URL in the mail must import it into their browser or an address book before any links can be used.

Bookmarks in software such as Gopher clients are specific to each client: Unless the client supports sharing of bookmarks and both the giver and receiver of information use the same client software, users will have to resort to the method of describing the path taken to the information. It is well within the capabilities of current browsers and protocols to provide explicit and convenient support for sharing information about Internet information. More sophisticated environments to support saving Internet information for future use, as outlined earlier, would clearly be able to play a role in sharing information as well.

*Structuring information for use by others.* The ability to post information that (anonymous) others could access began with FTP, where information was organized in a hierarchy of directories and files. Gopher servers are not hierarchical, but present information in essentially the same format: directories and files. The WorldWideWeb offers posters of information the ability to add verbal description and context to its hypermedia links and fetchable files.

Although each of these environments, in turn, has added richness, improving a poster's ability to communicate the nature of the information being posted, and to guide users into using it more effectively, the environments and protocols of the future must go further. With the plethora of information available over the Internet, it will be crucial to support this kind of "metacommunication," helping posters to convey to users what information will be found at a site, and critical aspects of its use—for example, the information owner's Acceptable Use Policy, or prerequisites for using the information (technological or otherwise).

**Handling Dangerous Information**

Thankfully, the majority of Internet users have probably been more exposed to unpleasant or dangerous consequences of Internet use through the media than by their actual use of the net. Yet the worries are real, and based on real, if infrequent, occurrences. The issues are diverse, and consequently we should expect the approaches to them to differ as well.

In general, we can categorize problematic information in the following way:

*Harmful*: Information that may do damage to a user's computing environment. Examples are viruses and other nasty programs designed to destroy information or property.

*Controversial*: Information that appeals to some people, outrages other people, and tends to cause philosophical debate on freedom of expression. Includes material such as pornography, tasteless jokes, political commentary, religious treatises.

*Untrue*: Information recorded somewhere, used for some purpose, that is incorrect. Classic examples include false personal information in databases (e.g., name and address, credit history, stolen property reports, etc.). Well-known Internet examples (LaQuey & Ryer, 1993) include the proliferation of a plea to help a dying boy get into the Guinness Book of World Records for receiving the most Get-Well cards ever, and the broadcast of a cookie recipe supposedly from a famous restaurant that charged a surprised customer $250 (who had thought the waiter said "$2.50").

*Private*: Information that the owner considers confidential, yet may be accessible to others. Examples include financial transactions executed over the net, what information a user has browsed and/or downloaded, and the content of e-mail messages.

*Stolen*: Information that has been deliberately removed from its intended context with an intent to abuse it. Examples are credit card numbers and passwords sent over the net that are captured and reconstructed by surreptitious programs at Internet way-stations.

A collection of approaches will be necessary to defend against these different problems of unwelcome information or information use. Virus detection programs on servers to prevent harmful code from being posted, better mechanisms for publicizing known dangers to Internet users, and improved ability to trace harmful information to its source and hold perpetrators responsible will improve protection from damaging programs. Encryption and other security schemes are becoming available to address the need for the confidentiality and security of financial transactions. The goals for dealing with harmful or stolen information are clear: Addressing problems in these arenas is a matter of placing effort into creating redundant, near-failsafe mechanisms.

The goals are less clear for how to handle controversial, untrue, or private information, and the issues are more complex. Clearly, for example, it would be nice if users could be alerted to possibly offensive material *before* they retrieve it—the way some television broadcasters now warn viewers before a program begins of sex, violence, foul language, or possibly distressing content. Equally obvious, it would be impossible to label every piece of Internet information around the world with some kind of "rating" with respect to its content. Instead, we expect to see Internet information handled largely via the same mechanisms used for other published information in the world, such as books and magazines. "Edited" collections will be created from particular points of view; indeed, these are already ubiquitous on the Internet. Particular groups might generate their own schemes for rating limited amounts of information (e.g., similar to granting a "good housekeeping seal of approval" on particular files or servers). In some cases, Internet technology may play a role; for example, enabling access to only particular servers in an educational setting, as the School Internet software is capable of doing.

Untrue information raises another set of concerns. At present there is little way to deal with outdated information that keeps circulating via the net—for example, the appeal to help Craig fulfill his dream of being in the Guinness Book of World Records. Currently, debunking and counteracting such messages before they generate a new flood of responses depends on someone who knows better seeing and responding quickly to the message. Perhaps in the future, ways to verify that a request is still valid can be built into the structure of messages.

Private information presents a different problem, particularly for information having both *personal* and *nonpersonal*, or *business*, aspects. As many consumers know, getting false (personal) information in a (privately owned) database corrected can be exceedingly difficult. Furthermore, information that users may consider private may become easily accessible (and shareable) by companies. It is not clear at present who "owns" such data (e.g., information about how much money a user has spent via the net, and on what products, what information has been browsed by which users)—the buyer or the seller— nor is it clear who has what rights to use such information, and in what ways. Hill and Hollan (1992) have raised this issue in the context of their work on document processors that record and display user interactions; specifically, which parts of a document users have edited ("Edit Wear") or read ("Read Wear"). As soon as interaction histories for various interface objects (e.g., documents) are permanently recorded, ethical issues are raised as to ownership and use of such data.

As the commercialization of the net and enhancements to its infrastructure take place, conventions and/or laws addressing these issues will emerge. Perhaps it will be the case, for example, that the user will be deemed to "own" information representing his or her individual behavior patterns, while companies "own" (and can share) aggregated information about transaction and/or information-usage patterns. There will, however, be gray areas: Is sending a targeted advertisement to individuals making up a relevant pattern a legitimate use of aggregated data? Advocacy organizations such as EFF (the Electronic Frontier Foundation) are playing an important role by articulating new, and sometimes hidden, issues being raised by the capabilities of Internet technology.

## HCI AND THE INTERNET

The analysis presented in the previous section organizes observations about Internet usefulness and usability in terms of fundamental information-using tasks faced by Internet users. In this section, we discuss the relationship between HCI design and Internet protocols based on our experience in designing a suite of Internet applications, and what HCI might be able to suggest about the future evolution of the protocols. We also attempt to abstract and encapsulate our recommendations based on the HCI principles and research results that have been brought to bear in the course of the earlier discussion. Finally, we attempt to ward off the conclusion that improvements in the performance of the infrastructure per se will be sufficient to improve Internet usability and usefulness.

### HCI Design and Internet Protocols

IBM School Internet is a suite of applications for Internet access designed primarily for teachers and students. In the course of designing and implementing School Internet, there was a constant interplay between design ideas and the possibilities for implementing them given our understanding of the underlying protocols. To illustrate this, we offer the following examples in the arenas of mail and newsgroups:

1. *Reading Mail.* A minimal requirement for a usable mail reader is to present a reasonable description of the user's mail. Such a description might include, at a minimum, the name of the sender, text from the subject line, and the date of arrival. Surprisingly, the most common server protocol for handling mail (POP for Post Office Protocol) makes creating this relatively onersome for a client application.

The reason lies in the nature of the interaction that a client application is allowed to have with the server. With regard to queries relevant to building a description of mail, clients can ask a minimally conforming POP server for either an entire mail item (consisting of header and body), or just the header. The header contains all the information needed to build the minimal description of mail outlined earlier, but the client must parse the header and extract the desired content before it can be displayed to the user. Moreover, the client must ask the server for this information for each individual mail item in turn (identified by a number which the client accesses by first issuing a "list" request to the server). The constraints of the protocol, in this case, mean that clients must do a large amount of processing to recoup (from an HCI perspective) minimal information, which has the further consequence of potential performance decrements.[5]

There were other design ideas in the course of our work that, like building a decent description of a user's mail, were probably possible, but not particularly easy or convenient, given the protocol. For example, we saw a need to handle e-mail generated by discussions such as LISTSERV's in a way separate from nonlist mail. Some lists generate a huge amount of mail, and users' mailboxes are swamped by it. Deleting large numbers of unread mail items is a common practice, judging by our own experience and statements we have read from many participants in such discussions. What is needed in order for e-mail discussions to be segregated is for information uniquely identifying the source as a LISTSERV to be present in the header.

The same story prevailed for other ideas, such as displaying status information for users about their mail (whether a mail item was newly arrived or old, and, among old items, whether they had yet been read). We would have liked to experiment with allowing users to manipulate mail items—for example, renaming or annotating them while remaining in the mail list. Once again, it is surely possible to implement any of these ideas with sufficient effort and storage on the client side, but the protocol is not designed to support the accumulation or addition of "history" for individual mail items, making the design and implementation of such function unwieldy and conceptually messy.

Limitations on client function imposed by underlying protocols are not lost on the Internet developer community, of course. Electronic mail, as one of the best-established functions of the Internet, is a particularly clear example of how the evolution of Internet protocols occurs. Many limitations of the POP protocol, for example, are beginning to be addressed by extensions to the protocol itself, by additional protocols that function alongside or in cooperation with the

---

[5]Our particular solution to this problem was to go outside the POP protocol (because we controlled the behavior of both the clients and the servers in our context). We extended our mail server protocol, modeling after the XHDR command in the NNTP News protocol, such that the client can ask for the content of particular header fields across a range of mail items (yielding already-parsed content from a single client request).

POP protocol (e.g., Remote Mail Checking), or by new protocols that replicate and extend the established protocol (e.g., IMAP: Interactive Mail Access Protocol). The standard for the format of Internet text messages has likewise been extended to accommodate multimedia, multipart messages by the MIME (Multipurpose Internet Mail Extension) protocol. We discuss protocol evolution further later.

2. *Usenet News.* Our example from the domain of Usenet News, and the NNTP protocol, is slightly different. Here is a case where the protocol anticipates a useful user function—identifying contributions to the same "thread" of conversation—but one that is rarely implemented in accordance with the protocol by clients that support reading and contributing to news discussion groups.

"Threads" can do much to enhance the usefulness of news discussions for users. Identifying threads allows news clients to implement function such that users can display a synopsis of the current topics under discussion in a newsgroup, or can ignore threads they find unappealing. It also enables users to address other potential actions of interest to the thread, such as printing or saving.

The standard for interchange of Usenet messages specifies that clients should offer a "follow-up" action to users who are reading a news article. When the user invokes that action, the protocol says that the client must provide appropriate reference information in a "References" header field attached to the user's follow-up post. The Reference field should contain the unique message ID carried by the article to which the follow-up is being generated (which is already carried by the article). Instead, many clients use the "Subject" field in the header to identify the referred-to article; a needlessly less accurate method of achieving similar functionality.[6]

A second issue that surfaces with usability impact in news readers is that of synchronization between client and server. A user who leaves his or her news client running can come back to it days later to find that most of the articles displayed in a newsgroup no longer exist on the server. Groups and/or articles can disappear from the server in much shorter time periods, too, of course, but "refreshing" the contents displayed by the client is, ideally, neither a user's responsibility, nor a batched operation. Unfortunately, given the current protocol, the "refresh" function is always batched, and is always the user's responsibility—either explicitly (as in providing a "refresh" function that reloads all newsgroups and articles), or implicitly (as in exiting and restarting the news application).

---

[6]Interestingly, one reason for this may be the structure of the standards document itself, which places the specification of this (required) function in a section of the document entitled "Optional Headers."

## General Implications for the Evolution of Protocols

As we remarked earlier, the accomplishments of the original architects of the Internet via the creation and publication of protocols are remarkable. We find no fault with them here; they succeeded in addressing difficult issues involved in constructing a robust infrastructure over a heterogeneous set of machines. However, as the Internet and its protocols develop in the future, we expect to see a shift in emphasis as issues arising from user needs and Internet usage are more thoroughly addressed by the protocols. The examples from the School Internet project given previously are only illustrative; throughout the design process there were myriad details and decisions that hung on our understanding of what was possible via the protocols. From this experience, the following points have emerged for us:

1. *Servers must become less passive with respect to their clients.* We pointed to the issue of synchronization between server and clients earlier. This is an issue that surfaces in more than one context, and can be seen as a consequence of the more general fact that the capabilities and roles played by servers and clients are asymmetric. Clients ask; servers respond. Servers do not generally speak until spoken to, nor do they "alert" clients to potentially interesting information.

2. *Access to server information must transcend the "all-or-none" mode.* A point not unrelated to the first, clients must be able to ask servers more sophisticated questions about the information they hold, and be able to retrieve partial information rather than undifferentiated chunks (as in the earlier mail example).

3. *Protocols must better recognize and support user needs.* Hand in hand with the ability to request partial information from servers, we expect the kind of information (optionally) associated with Internet objects such as files and directories to expand in recognition of commonly needed "metainformation." We have discussed the need for users to evaluate information via preview mechanisms. Descriptive and other sorts of information (copyright, acceptable use) might be attached to Internet objects in the future, and supported by the servers that hold and present them.

A useful exercise for discovering some of the ways Internet protocols might (and should) evolve in the future is to look at the kinds of complaints, errors, and "work arounds" that Internet users exhibit now. When we engaged in this exercise, we noted many promising candidates for future protocol support, for example: automatic redirection of subscribe/unsubscribe messages mistakenly sent to a discussion list rather than its administrative address, tagging of "special" kinds of information such as FAQ files, discussion group "purpose" and "rules," and administrative information about lists (e.g., how to stop delivery while a user is on vacation).

## A Recap of HCI Recommendations for the Internet

In the course of this chapter, we have made several recommendations for addressing usability and usefulness issues for Internet users, based on well-established HCI principles and HCI research results. Here we distill these ideas into a few more broadly stated recommendations.

1. *Support the stages of user action.* Norman (1986) presented an analysis of seven distinct stages of user activity, including Establishing a Goal, Forming an Intention, Specifying an Action Sequence, Executing the Action, Perceiving the System State, Interpreting the State, and Evaluating the System State with respect to the Goals and Intentions. Many of the usability problems discussed in this chapter can be viewed as inadequacies in support for one or more of these stages. For example, helping users to evaluate information they might fetch (via preview or some other mechanism) supports Establishing Goals that are meaningful and therefore not a waste of the user's time or Internet bandwidth.

2. *Support the repair of errors.* In an ideal world of HCI, user goals are always accomplished without much ado or errors. In the real world, errors are commonplace, whether the "credit" can be assigned to the user, the system, or the interaction between them. Anticipating and designing for error, and error recovery, are hallmarks of the best design. Carroll, Kellogg, and Rosson (1991), Lewis and Norman (1986) and Norman (1983) offered suggestions on designing for error.

3. *Consider the context of information use for users.* Compatibility with the user's context of use, and contributing to the user's larger work or personal goals are the primary yardsticks by which users measure the usefulness of Internet tools. Sproull and Kiesler (1986) and Kraut, Galegher, and Egido (1988) provided illuminating examples of considering the relationship between human activity and technology for its support. Ehn (1988) provided an in-depth discussion of this issue.

4. *Exploit techniques from relevant HCI literature and research.* Many topics pertinent to the human factors of Internet use have been extensively explored in other contexts. Useful techniques exist in literature addressing such topics as information retrieval (e.g., Dumais, 1988), hypertext and navigation of complex spaces (e.g., Nielsen, 1990), visualization of complex information (e.g., Card, Robertson, & Mackinlay, 1991), and Computer-Supported Cooperative Work (CSCW; e.g., Greif, 1988).

5. *Enhance the user's ability to create and manage a "personal" world* of Internet information. Internet tools are more than a means of navigating cyberspace; they are also a primary basis on which users conceptualize and preserve the information of most value to them. The ability to provide flexible and effective means for users to mark, annotate, maintain, and manipulate information and pointers to information will increasingly differentiate the most useful tools.

Jones (1988) and Erickson and Salomon (1991) provided interesting starting points for thinking about this issue.

6. *Think "object-oriented."* Several of the changes suggested in this chapter suggest a need for the Internet to function in a more object-oriented way. Many "objects" of the Internet are now sufficiently well-established to consider enhancing their ability to behave or support queries—for example, mail messages and files. Some of the ideas we have offered along these lines are pointers that update themselves in the user's environment, messages that enable automated checks of the (current) validity of their requests, and information (voluntarily) associated with files that allows users to be warned of potentially offensive material before they view it.

**Why "More Bandwidth" Is Necessary, But Not Sufficient**

We want to be clear: We eagerly anticipate the day that an incredibly inexpensive T1 line to our private residences is available. Bandwidth is good. Higher quality infrastructure is good. However, none (or very few) of the human factors issues outlined here can be addressed with these kinds of "bigger and better" technical solutions. Getting the wrong information, or not enough information, faster is not a solution, only (at best) less of a problem. The most challenging issues of Internet usability and usefulness are pragmatic and social in nature: For example, how do users capitalize on the Internet as an information resource within their own constraints of space and time? How will information ownership, or value added to information, be determined and accounted for on the Internet? How can information providers make their information most useful and attractive to others?

The pioneering developers of the Internet and its protocols have accomplished amazing feats; that the Internet functions as well as it does under the stress of its current rate of growth is ample testimony to that. As the Internet passes through its next phase of evolution, we trust we will see ever greater realization of its ability to support the human use of information as discussed herein.

## SUMMARY

The Internet is unlike any previous communications technology deployed on a world-wide scale. It affords one-to-many communications that can be rich in bandwidth and interactive. The infrastructure, protocols, tools, communities, conventions, and activities of the Internet are evolving simultaneously and rapidly.

Landauer (1995) drew a distinction between what he described as two partially overlapping phases of evolution in the deployment of information technology, which, not coincidentally, are largely associated with different types of applications. The original application of information technology was to automate functions for which humans are not particularly well suited (such as rapid numerical operations). The second phase of evolution, in contrast, has focused on applications that attempt to enhance and extend human ability in domains and tasks that cannot be taken over completely by numerical machines—a type of technology he referred to as "augmentation." Landauer provided a compelling argument that augmentation technologies cannot succeed in making users more productive unless they are designed using well-established cognitive and usability engineering principles and practices. The perils of ignoring the users and their context of use, or of pursuing purely technical solutions ("all that's needed is more bandwidth") to what are essentially human and social problems, have been noted again and again in the HCI literature. The human factors of the Internet can only become more critical as its population of users expands to include less technically oriented people and as their use expands to near continual contact with the net.

With the advent of Internet readiness as a standard part of operating systems, the Internet is about to enter an era in which weaknesses in human–computer interaction could begin to threaten its viability. This prospect only increases the importance of examining Internet technology from the point of view of users and improving its ability to support human needs and activities. The urgency and broad scope of the challenge to make the Internet useful and usable is fine with us. From where we sit, making users of the Internet productive and happy while continuing to envision and enable the possibilities of world-wide networking is the most interesting HCI challenge of the next decade.

## REFERENCES

Ahlberg, C., Williamson, C., & Shneiderman, B. (1992). Dynamic queries for information exploration: An implementation and evaluation. In P. Bauersfeld, J. Bennett, & G. Lynch (Eds.), *Human factors in computing systems: The proceedings of CHI'92* (pp. 619–626). New York: ACM Press.

Bannon, L. J. (1986). Helping users help each other. In D. A. Norman & S. W. Draper (Eds.), *User-centered system design: New perspectives on human-computer interaction*. Hillsdale, NJ: Erlbaum.

Barsalou, L. W. (1983). (1983). Ad hoc categories. *Memory & Cognition, 11*(3), 211–227.

Brothers, L., Hollan, J., Nielsen, J., Stornetta, S., Abney, S., Furnas, G., & Littman, M. (1992). Supporting informal communication via ephemeral interest groups. In J. Turner & R. Kraut (Eds.), *Proceedings of the Conference on Computer-Supported Cooperative Work* (pp. 84–90). New York: ACM Press.

Calcari, S. (1994, September). A snapshot of the Internet. *Internet World*, pp. 54–58.

Card, S. K., Moran, T., & Newell, A. (1983). *The psychology of human- computer interaction*. Hillsdale, NJ: Erlbaum.

Card, S. K., Robertson, G. G., & Mackinlay, J. D. (1991). The information visualizer: An information workspace. In S. Robertson, G. Olson, & J. Olson (Eds.), *Human factors in computing systems: The proceedings of CHI'91* (pp. 49-54). New York: ACM Press.

Carroll, J. M., & Kellogg, W. A. (1989). Artifact as theory-nexus: Hermeneutics meets theory-based design. In K. Bice & C.H. Lewis (Eds.), *Human factors in computing systems: The proceedings of CHI'89* (pp. 7–14). New York: ACM Press.

Carroll, J. M., Kellogg, W. A., & Rosson, M. B. (1991). The task-artifact cycle. In J. M. Carroll (Ed.), *Designing interaction: Psychology at the human–computer interface* (pp. 74–102). Cambridge, MA: Cambridge University Press.

Clinton, B., & Gore, A. (1992). *Putting people first: How we can all change America.* New York: Times Books.

Clement, G. P. (1994, September). Library without walls. *Internet World, 60–64.*

Cronin, M. J. (1994). *Doing business on the Internet.: How the electronic highway is transforming American companies.* New York: Van Nostrand Reinhold.

Dern, D. (1994). *The Internet guide for new users.* New York: McGraw-Hill.

du Boulay, B., & Matthew, I. (1984). Fatal error in pass zero: How not to confuse novices. *Behaviour & Information Technology, 3,* 109–118.

Dumais, S. (1988). Textual information retrieval. In M. Helander (Ed.), *Handbook of human–computer interaction.* Amsterdam: Elsevier Science.

Ehn, P. (1988). *Work-oriented design of computer artifacts.* Stockholm, Sweden: Arbetlivscentrum.

Erickson, T., & Salomon, G. (1991). Designing a desktop information system: Observations and issues. In S. Robertson, G. Olson, & J. Olson (Eds.), *Human factors in computing systems: The proceedings of CHI'91* (pp. 49–54). New York, NY: ACM Press.

Furnas , G. W., & Zacks, J. (1994). Multitrees: Enriching and reusing hierarchical structure. In B. Adelson, S. Dumais, & J. Olson (Eds.), *Human factors in computing systems: The proceedings of CHI'94* (pp. 330–336). New York: ACM Press.

Goldstein, J., & Roth, S. F. (1994). Using aggregation and dynamic queries for exploring large data sets. In B. Adelson, S. Dumais, & J. Olson (Eds.), *Human factors in computing systems: The proceedings of CHI'94* (pp. 23–29). New York: ACM Press.

Goodman, S. E., Press, L. I., Ruth, S. R., & Rutkowski, A. M. (1994). The global diffusion of the Internet: Patterns and problems. *Communications of the ACM, 37*(8), 27–31.

Greif, I. (Ed.). (1988). *Computer-supported cooperative work: A book of readings.* San Mateo, CA: Morgan Kaufmann.

Hill, W. C., & Hollan, J. D. (1992). Edit wear and read wear. In P. Bauersfeld, J. Bennett, and G. Lynch (Eds.), *Human factors in computing systems: The proceedings of CHI'92* (pp. 3–9). New York: ACM Press.

Jones, W. P. (1988). "As we may think"?: Psychological considerations in the design of a personal filing system. In R. Guindon (Ed.), *Cognitive science and its applications for human–computer interaction.* (pp. 235–287). Hillsdale, NJ: Erlbaum.

Kantor, A., & Neubarth, M. (1994, October). Internet interfaces: The next generation. *Internet World,* pp. 30–32.

Kellogg, W. A., Carroll, J. M., & Richards, J. T. (1992). Making reality a cyberspace. In M. Benedikt (Ed.), *Cyberspace: First steps* (pp. 411–431). Cambridge, MA: MIT Press.

Kraut, R., Galegher, J., & Egido, C. (1988). Relationships and tasks in scientific research collaborations. *Human-Computer Interaction, 3*(1), 31–58.

Landauer, T. K. (1991). Let's get real: A position paper on the role of cognitive psychology in the design of humanly useful and usable systems. In J. M. Carroll (Ed.), *Designing interaction: Psychology at the human–computer interface* (pp. 60–73). Cambridge, U.K.: Cambridge University Press.

Landauer, T. K. (1995). *The trouble with computers: Usefulness, usability, and productivity.* Cambridge, MA: MIT Press.

LaQuey, T., & Ryer, J. C. (1993). *The Internet companion: A beginner's guide to global networking.* Reading, MA: Addison-Wesley.

Lewis, C., & Norman, D. (1986). Designing for error. In D. A. Norman & S. Draper (Eds.), *User-centered system design: New perspectives on human–computer interaction.* Hillsdale, NJ: Erlbaum.

Minsky, M. (1961). Steps towards artificial intelligence. *Proceedings of the Institute of Radio Engineers, 49,* 8-30.

Miyata, Y., & Norman, D. A. (1986). Psychological issues in support of multiple activities. In D. A. Norman & S. Draper (Eds.), *User-centered system design: New perspectives on human–computer interaction* (pp. 265–284). Hillsdale, NJ: Erlbaum.

Murray, J. (1993). K–12 Network: Global education through telecommunications. *Communications of the ACM, 36*(8), 36–41.

Nielsen, J. (1990). *Hypertext and hypermedia.* Boston: Academic Press.

Norman, D. A. (1983). Design rules based on analysis of human error. *Communications of the ACM, 26*(4), 254–258.

Norman, D. A. (1986). Cognitive engineering. In D. A. Norman & S. W. Draper (Eds.), *User-centered system design: New perspectives on human–computer interaction* (pp. 31–61). Hillsdale, NJ: Erlbaum.

Norman, D. A. (1992). *Turn signals are the facial expressions of automobiles.* Cambridge, MA: Addison-Wesley.

Rheingold, H. (1993). *The virtual community: Homesteading on the electronic frontier.* Reading, MA: Addison-Wesley.

Schank, R. C. (1982). *Dynamic memory.* New York: Cambridge University Press.

Schank, R. C., Collins, G. C., & Hunter, L. E. (1986). Transcending inductive category formation in learning. *The Behavioral and Brain Sciences, 9*(4), 639–651.

Shneiderman, B. (1980). *Software psychology: Human factors in computer and information systems.* Cambridge, MA: Winthrop.

Soloway, E. (Ed.). (1993). Technology in K–12 Education. *Communications of the ACM, 36*(5), 28–30.

Sproull, L., & Kiesler, S. (1986). Reducing social context cues: Electronic mail in organizational communication. *Management Science, 32*(11), 1492–1512.

Styx, G. (1993). Domesticating cyberspace. *Scientific American, 269*(2), 100–110.

# Chapter 2
# Computer-Augmented Environments: New Places to Learn, Work, and Play

Benjamin B. Bederson*
*Bell Communications Research*
*Computer Graphics and*
*Interactive Media Research Group*

Allison Druin
*New York University*
*Computer Science Department*
*Media Research Laboratory*

## INTRODUCTION

People need to communicate with one another—they seek to express themselves in new creative ways and look for new avenues to learn, work, and play. We have come to augment our activities with technology because it enhances our human capabilities. For the past 30 years these activities have increasingly included large numbers of plastic computer boxes with keyboards, mice, and displays, the promise being that these desktop technologies could promote more efficient and better quality communication and interactions with each other.

To some extent, this promise has come true thanks to recent developments that provide inexpensive, miniaturized, and more powerful technologies. This can be seen in everything from electronic mail to international banking to telephone networks. In a different light, however, one could say that technology has also served to isolate us from each other and our physical surroundings. Our social interactions are no longer direct, but instead are mediated, as with e-mail (keyboards and text), phone calls (telephone receivers and sound), and banking (touch screens and keypads).

It is our belief that people still want and will continue to need face-to-face interaction in shared social spaces. For example, people still choose to attend music concerts, even though listening to a compact disc at home can be more convenient and the sound quality is perhaps better. People also still choose to discuss business in social settings such as restaurants, instead of taking advan-

---

* Benjamin B. Bederson has moved to the University of New Mexico where he can be reached at: bederson@cs.unm.edu.

tage of less expensive methods of communication, such as meeting in a conference room, or even e-mail.

Today, researchers are looking beyond the desktop box to computer-augmented environments, which take advantage of physical devices, spaces, and media to augment and enhance our interactions with people and objects. The confines of computer augmented environments are not keyboards, mice, or data gloves; rather, they extend to the walls in which we live, the furniture in which we sit, and the common objects with which we interact. Computer-augmented environments don't attempt to replace the real world. Instead, they add to it—taking advantage of the complexity of the world and the ability of computers to add to it.

A growing number of researchers have begun to focus their energies on computer-augmented environments, a broad term covering a spectrum of technologies as mundane but important as supermarket barcode scanners to technologies as exciting and novel as electronic paper. This was not the case in the recent past. Researchers seemed to be more concerned with creating completely simulated virtual realities. With current technology, however, people have found it difficult to simulate the detail and richness of the real world in real time. By augmenting rather than replacing the world, researchers have found computer-augmented environments to be a powerful combination of sophisticated technology and rich physical spaces.

The use of computers to enhance the physical world goes by many names including "augmented reality," "artificial reality," and "immersive environments." A pivotal summary of this work was highlighted in July 1993 in the *Communications of the ACM*, which contained 12 articles from researchers throughout the world (Wellner, 1993). This compilation of work was called "Computer-Augmented Environments: Back to the Real World". In recognition of this publication, we have also chosen to refer to this chapter as "Computer-Augmented Environments." However, we see two distinct classifications of these environments:

1. Information superimposed on the physical world
2. Information technologies integrated or embedded into the physical world

In the first case, information is superimposed on the physical world via such devices as audio headphones or see-through, head-mounted displays. The purpose of these environments is not to replace what one sees or hears in the physical world, but to add to what is already there. In the second classification of computer-augmented environments, various sensor and output technologies are embedded into the physical world. When sensors are triggered by a person's movement, touch, or sound, information can be offered through changes in the environment, sound, video, lighting, and so on.

This chapter discusses numerous examples of computer-augmented environments, their supporting technologies, historical roots, and future possibilities. The examples we have chosen to highlight are by no means an exhaustive list of all computer-augmented environments; rather, they are selected examples the authors feel represent a cross-section of work in the field.

## HISTORICAL PERSPECTIVE

Although today's computer-augmented environments introduce new technologies and design methodologies, they are also deeply rooted in the past. For every startling "new" discovery of today, there are usually several well-documented research projects that one can point to in the past that have had a strong influence on the current work. Physical computer environments have been evolving at least since the late 1960s, when researchers started to redefine *how* and *where* people physically interact with technology.

One of the first demonstrations of a computer-augmented environment was given by Sutherland in 1968, who, for the first time at an international conference, presented a head-mounted, three-dimensional display (Sutherland, 1968). The first applications of this technology came in the aerospace industry, where heads-up displays were used to add computer-generated graphics to the scene a pilot saw through the windshield of a fighter jet (Furness, 1969).

In the areas of education and entertainment, Krueger (1991, 1993) was one of the first researchers to explore full-body interaction with technology in a room-sized space. For over two decades, Krueger has focused on "computer controlled responsive environments." He has held a strong belief that the ultimate computer interface is the human body. Since 1970, Krueger has been creating what he has come to call "artificial reality" with his research on "VIDEOPLACE" (see Figure 1). In this work, the whole human body becomes an input device, and room-sized screens show the output. He sees applications for this work in such areas as shared work spaces, scientific exploration for children, and physical therapy. Although this work does serve to replace most of our physical environment with an artificial one, VIDEOPLACE has exemplified what can be done with fully interactive technology.

The research of MIT's Architecture Machine Group (a precursor to the MIT Media Lab) also explored room-sized, computer-controlled environments (Bolt, 1984). In the MIT "Media Room," people would sit in a chair and verbally "tell" and physically point to what they would like moved or changed on wall-sized screens. This research, led by Bolt and Negroponte, was groundbreaking in its philosophy. This group did not only believe in full physical interaction, but also in the quality of the physical environments that technology could enhance.

Figure 1: Myron Krueger's VIDEOPLACE.

Bolt's and Negroponte's philosophy can perhaps best be summarized thus:

> The human interface with computers is the physical, sensory, and intellectual space that lies between computers and ourselves. Like any place this space can be unfamiliar, cold and unwelcoming. But it can also be like some other places, those we know and love, those that are familiar, comfortable, warm and most importantly personal. (Bolt, 1984, p. xiii)

Research that has invited interaction with rich physical spaces can also be found in children's computational environments. Since the 1960s, Papert (1980) has lead a team of researchers at MIT in developing Logo, a children's programming language, culture, and learning environment. The most popular use of Logo in the 1970s involved a "Floor Turtle" (see Figure 2). It was a simple mechanical robot connected to the computer. Children could program in Logo to make the turtle go forward, backward, left, or right. In this way the physical turtle became a concrete way for children to use what they understood about the physical world to better understand geometry. The turtle did not replace their physical surroundings; rather, it was part of their learning environment. Children could be found imagining themselves to be the turtle, to better understand how to program their environment.

By the 1980s, computer-enhanced environments for children also included NOOBIE, a 5-foot, Muppet-like creature with fur, feathers, and an iridescent fish tail (Druin, 1988). This research, led by Druin as a part of Apple's Vivarium Research Group at the MIT Media Lab, was a new approach to designing an animal design playstation. Druin asked children to interact with a computer in a way intuitive to them—by hugging a stuffed animal (see Figure 3). Instead of using a mouse or keyboard children squeezed the different parts of NOOBIE to create fantasy creatures on the screen housed in NOOBIE's belly. For example, by squeezing NOOBIE's nose, a child would hear a sound and see a nose on the screen.

Since the mid-1980s, Druin has continued to embed computer technologies in familiar objects and places for children. Some of her work has included squeezable stuffed globes, books, and clothing.

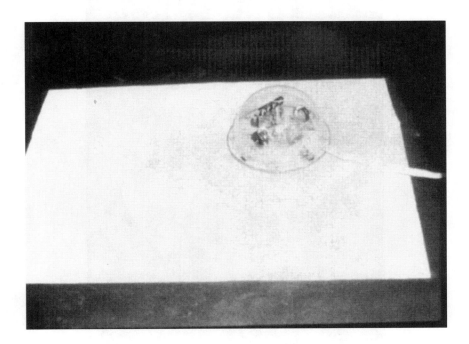

Figure 2: Seymour Papert's Floor Turtle.

## EXAMPLES OF COMPUTER-AUGMENTED ENVIRONMENTS

In this section, we describe several examples of recent computer-augmented environment research. We have categorized these projects into two groups:

1. Information superimposed on the physical world.
2. Embedding information in the physical world.

### Information Superimposed on the Physical World

The group of projects that follow all superimpose information on the physical world. By having people carry the necessary equipment with them, their sight or

Figure 3: Allison Druin's NOOBIE, an interactive animal design playstation.

hearing can be augmented with computer-generated data. This group of projects is similar to the popular notion of virtual reality. However, in this case all the projects described augment, rather than replace, the physical world. The first three add images to the environment, whereas the last one adds sound.

*Knowledge-based augmented reality—Steve Feiner et al., Columbia University*. One promising application of visually augmented reality is helping people handle complex three-dimensional tasks. By highlighting, labeling, and animating relevant objects, such systems could simplify tasks by showing what to do at the place of action. Traditionally, such visual support could be found only on a piece of paper with a diagram of the real objects. This leaves it to the viewer to match the diagram to the physical object.

Designing three-dimensional augmented reality information presentations can be quite complicated. Such systems must understand the information being presented well enough to always show the appropriate images. This becomes more difficult with increasing richness of information. For instance, in three dimensions, informative labels must be attached to objects so that they are always visible, despite the fact that the viewer can look at an object from any viewpoint. The system designed by a group led by Feiner (Feiner, MacIntyre, & Seligmann, 1993) at Columbia University addressed these issues by creating a knowledge-based augmented reality system.

Their project, called *Knowledge-based Augmented Reality for Maintenance Assistance* (KARMA) is built on top of the Intent-Based Illustration System (IBIS) (Seligmann & Feiner, 1991). The idea is to specify a prioritized list of presentation goals, such as showing the location of an object. IBIS is a rule-based system that examines potential solutions and backtracks when errors occur. The presentations are generated dynamically; as parameters (such as viewpoint) change, IBIS monitors the scenario and changes the illustration as is appropriate. For example, an outline of an object may be designated unoccludable (always visible). In this case, as the viewpoint changes, if the outline becomes blocked by another label or graphic, the blocking graphic may be deleted, rendered partially transparent, or rendered with a cutaway view.

The application of laser printer maintenance was used to test the system. The goal was to visually assist the user in refilling the paper tray and replacing the toner cartridge. For each task, outlines of the relevant objects are superimposed on the laser printer and the appropriate actions are indicated (see Figure 4).

Several pieces of tracking and display technology are used in this and other augmented-reality systems to create the full immersive experience. (See the section on Supporting Technologies for more information.)

In this example, Feiner's group used a display device that is viewed through a half-silvered mirror. In this way, graphics are annotated on the physical world, so it is a true augmented system. This system was implemented with several different processors operating in a loosely coupled parallel design, which enables real-time tracking of the viewer and the laser printer components while IBIS calculates how to present the information. In this way, images are dynamically rendered in real time for the viewer as they work.

*Augmented manufacturing processes—Thomas Caudell, Boeing Computer Services.* The manufacturing of large jet airplanes lends itself to the use of visually augmented reality. During the assembly process, which often requires a high level of manual dexterity and skill, workers frequently need to reference technical information that guides or describes the process at hand. For instance, a drawing may specify where a hole should be drilled on a metal part. The worker must have the drawing at hand, correlate it to the physical part as a guide.

With visually augmented reality, the hope is that a worker can wear goggles that superimpose images directly on the relevant parts. In this way, the computer-generated image can remain registered on the physical parts even as the viewer walks around and moves his or her head. An arrow pointing to the correct location could be superimposed on the part, specifying precisely where

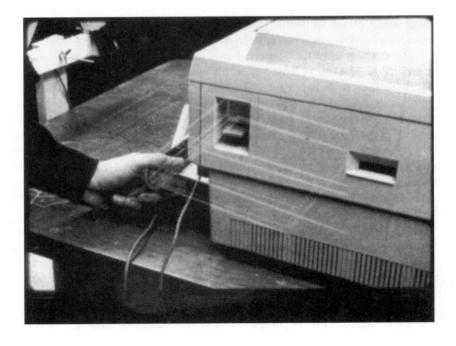

Figure 4: Steve Feiner's visually augmented reality to guide laser printer mainte-nance. (Copyright 1993 by Steven Feiner, Blair MacIntyre, & Doree Seligmann, Columbia University. Reprinted by permission.)

to drill a hole.

A research project at Boeing Computer Services (Caudell & Muzzle, 1992) is working toward this vision. They have developed a prototype to experiment with computer-augmented manufacturing processes and have developed four prototype applications, all of which operate on a fixed platform (see Figure 5). One application is for bundling wires to be used in airplane manufacturing. Currently, wire bundles are manufactured by using a computer-generated diagram glued to sheets of plywood. Pegs are placed in the plywood over the diagram and the wires are placed around the pegs. With visually augmented reality, a blank peg board is used and the computer generates the position of the pegs for the user to install. The paths of the wires are then indicated. With this system, thousands of formboards for wire bundles could be eliminated, and the markings on the formboard do not get covered up as the wires are layed out. Other applications use the same platform, and help with assembling wiring connectors, composite cloth layup, and physical parts for airplanes.

Their display system consists of a Private Eye™ (Reflection Technology, 1994) display mounted in front of one eye with a Polhemus™ (Polhemus Technology, 1994) head tracker all connected to a 386-based microcomputer system. The display occludes a small part of the field of view, but viewers can merge the images from both eyes to create a single augmented image (although some viewers have trouble with this).

As with all augmented-reality systems, calibration is a difficult practical problem because the computer generated images must be registered with the physical objects being viewed. This system is calibrated by adjusting the attachment of the display to the viewer while the display is in a fixed position relative to the work surface. Once the display is removed from the calibration position, it must not be moved relative to the viewers head, or it will have to be recalibrated.

***Situated information spaces—George Fitzmaurice, University of Toronto.*** There are several methods for augmenting the physical world with computer-generated information. Some superimpose visual or auditory information on the physical world with special glasses and headphones. Another approach, taken by Fitzmaurice (1993), is for the user to carry a small display in his or her hand that knows where it is.

This approach, which Fitzmaurice called *Situated Information Spaces*, enables the user to access information by retrieving it from the spatial location in which it was created. He gives the example of a fax machine, where a user could retrieve a log of incoming calls by moving the hand-held display near the speaker of the phone, and seeing the log on the display. Similarly, one could see other pieces of relevant information by moving the display near other points of the fax machine.

He has begun to explore this approach with a prototype device called the *Chameleon* (see Figure 6). It consists of hand-held monitor connected to the video display of a Silicon Graphics (SGI) computer. The portable monitor has a

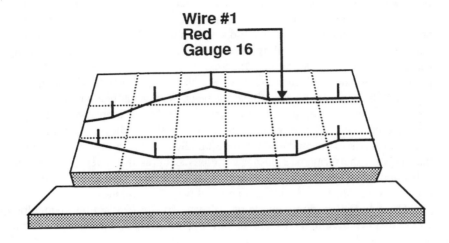

**Wire #1**
**Red**
**Gauge 16**

Figure 5: Thomas Caudell's platform-based augmented reality demonstration applications. This shows the wiring formboard. (Adapted from Caudell & Muzzle, 1992.)

spatial location sensor on it that allows the SGI to know where the monitor is within a 3-foot cube. In this manner, the SGI can generate different displays depending on where the monitor is held.

They have used this system for an application that enables users to access geographical data. By moving the hand-held monitor around a wall-sized map of North America, information about a particular city is given when the monitor is directly in front of it.

*Automated tour guides—Ben Bederson, Bellcore/NYU Media Research Lab.*[1] Ben Bederson has been exploring the use of audio-augmented reality as a method for automated tour guides for museums (Bederson, Davenport, Druin, & Maes, 1994). The standard taped tour guides that are common in museums today are also a type of audio-augmented reality; however, they are limited in numerous ways.

The standard taped tours are linear and preplanned, leaving the viewer little

---

[1]Thanks to Jim Hollan for many of the discussions at the start of this project, and to Chris Calabrese who developed much of the software for this system.

Figure 6: George Fitzmaurice's "situated information space."

choice in how to visit the exhibit. In addition, because the entire tour is of a limited duration, typically 45 minutes or so, only a relatively small subset of the
pieces on exhibit are actually described. Furthermore, the selection of pieces
described on the tour is chosen by the curator, and may not match the interests
of any given viewer.

The taped tour guides also interfere with some of the social aspects of visit-
ing a museum because it is very easy to get out of sync with a friend. It is not
uncommon for two people to go to the museum together, but because one person gets 30 seconds ahead of the other on the tape, they never see each other.
Because the viewer is continuously listening to the tape, he or she is shut out
from the environment and less likely to meet other people and have a group
experience.

The system being developed by Bederson combines digital audio, a microcontroller, and a spatial locating device that is used to play the description of the
piece the viewer is standing in front of. The effect is that the viewer can visit the
museum in any order he or she desires, and can hear descriptions of pieces simply by walking in front of them. When the viewer walks away from a piece, the
description stops.

In addition, because the entire tour is mediated by a computer, it is possible to automatically keep track of the interaction history of the viewer and the exhibit. With this information, it will be possible to customize the tour based on which pieces the viewer sees, and in what order. For instance, the system could relate the piece currently being described to a piece previously seen by a viewer during the same tour. Furthermore, it could make educated guesses about the types of pieces in which the viewer is interested and offer further information about them. Because the exhibit interaction history is being recorded by the computer, it could be given (anonymously) to the curator to give an accurate picture of how the exhibit is being viewed. There could even be a "restroom button" on the device that would give the viewer directions to the nearest restroom.

The technology behind this system integrates several existing technologies (see Figure 7). Small infrared transmitters are placed on the ceiling above each piece of art. Each transmitter sends a unique ID once a second. The viewer carries a computer-controlled digital audio source, a microcontroller, and an infrared receiver that is mounted in the headphones. The microcontroller knows where the viewer is by the ID that is received, and then controls the digital audio source to play the appropriate prerecorded track that describes the piece.

## Embedding Information into the Physical World

The following projects all place computational resources directly in the physical environment. The first two do so with a more traditional notion of computers, whereas the Ubiquitous Computing project examines what happens when computers can move or follow you within the environment, and the DigitalDesk project addresses the limits of paper. The last two projects described in this section look at how sophisticated technologies can be integrated seamlessly into our everyday living spaces (LEGO/Logo and Immersive Rooms).

*Ubiquitous computing—Mark Weiser, Xerox PARC.* One of the first groups to explore the concept of integrating wireless networked computers into the physical environment was led by Weiser (1993) at Xerox PARC. In 1988, they began to discuss the concept of *ubiquitous computing*. Like many other researchers described in this chapter, Weiser wanted to *integrate* the computer into our everyday activities, his idea being that your computational environment should follow you around. If you walk into the next office, the documents you are working on will show up on the available displays in the new office.

They approached this goal by building three different-sized computers that could be imbedded into the environment (see Figure 8). They are all pen-based and can communicate with each other wirelessly. The three sizes were aimed to match different kinds of interactions with information. The smallest one, the *PARCTab*, is analogous to Post-It notes. Fitting easily in the hand, they are designed to be truly ubiquitous, reaching the far corners of the environment.

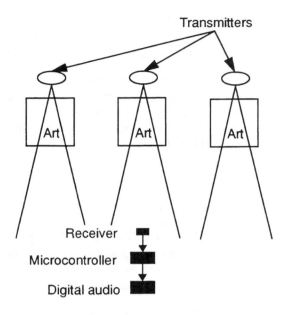

Figure 7: Schematic of Bederson and Calabrese's system for automated tour guides.

They are designed to store simple notes and provide basic search and retrieval functions. The medium-sized computer, the *MPad*, is notebook-sized. These are meant to be used like scratchpads with many being used by one person simultaneously. The largest computer, the *LiveBoard*, is wall-sized, similar to an office whiteboard. The LiveBoard, which is now a Xerox product, is well-suited to collaborative work or brainstorming within a large workspace.

With these and other technologies, the Xerox PARC group has begun to explore several applications. In using the LiveBoard, several tools were developed to explore collaborative shared work spaces—these tools allow physically separate users to work together using the LiveBoard and emphasizes pen-based drawing while accepting scanned input. It can also print the results and operate with several users working simultaneously on the same or different pages. Shared drawing tools are being explored at several places. While the concept is exciting, several issues are still being addressed. The ideal interface is still evolving—how it will scale up to support larger groups of simultaneous users is still a question.

Another application is based on the use of active badge technology pioneered at Olivetti Research Labs (Want, Hopper, Falcao, & Gibbons, 1992) to locate people and objects in the environment. It is similar to Bederson's work described previously, except that this work turns the technology around—it puts the transmitter on the person and the receiver in the room. In this system, a centralized computer tracks the occupants of the building, automatically forwarding phone calls and provides a map of the building to aid in finding its occupants and identifying spontaneous gatherings.

Similar technology is being used in the Responsive Environment project at Xerox PARC led by Elrod. This project automatically adjusts the local building environment (heat, light, and power) to accommodate the occupants. Simultaneously saving energy and providing a more comfortable work environment, this project demonstrates the type of things a truly invisible computer might do.

*DigitalDesk—Pierre Wellner, EuroParc.*[2] With the introduction of computers came the prediction of the demise of paper. Not only has the use of paper failed to decrease, but its use has thrived. Although the digital technology we use for simulated paper has continued to advance, the technology of real paper has not changed in a very long time. The *DigitalDesk* project attempts to improve the technology of paper by adding computer power directly to physical paper (Wellner, 1993). Operations on paper and computers each have their

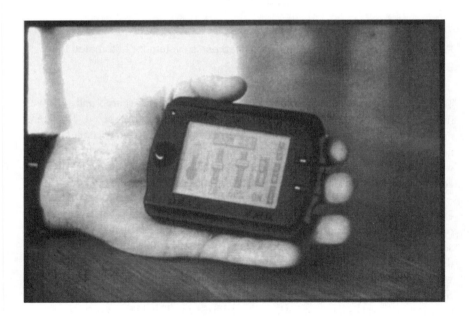

Figure 8: Mark Weiser's Ubiquitous Computing project at Xerox PARC: the PARCTab.

[2]Currently at AT&T Bell Labs.

advantages. For instance, searching is easier electronically, whereas reading is easier with paper. This project attempts to bring some of the power of computers to paper rather than the other way around.

The DigitalDesk project uses real paper on a real desk, but provides for electronic interaction with the paper by projecting computer images directly onto the paper and reading paper documents placed on the desk. It responds to pens or bare fingers, using a projector and camera placed above the desk and a sophisticated image processing system that can read paper documents and track finger motion.

There are three prototype applications for the DigitalDesk: a calculator, a drawing program, and a shared workspace. With common calculators, people often enter numbers that are printed on paper, and then copy the result back onto paper. The DigitalDesk calculator is projected onto the desk, and the user can copy a number from a paper on the desk to the calculator or vice versa with a simple hand/finger gesture. The calculator can thus be used as if it were standard—the system can detect the user tapping the projected buttons with a microphone mounted underneath the desk.

The painting program (PaperPaint) enables a person to draw real drawings on real paper, and then electronically copy and drag around any region of the drawing. Thus the image seen on the paper is a combined image of the projected electronic copies and the original penstrokes.

The shared workspace program (DoubleDigitalDesk) is reminiscent of Krueger's (1991) VIDEOPLACE, in which one user sitting at a desk draws on a piece of a paper while a second, remote user draws on paper on his or her desk. Each person sees the projection of the other on their desk, superimposed on their paper. It is therefore possible to play a game of tic-tac-toe by drawing on a real piece of paper at two separate desks.

*LEGO/Logo—Mitchel Resnick, MIT Media Lab.* Logo research is ongoing at the MIT Media Lab. Today the mechanical Logo turtle has been replaced by "LEGO/Logo" and "Electronic Bricks" (Resnick, 1993; see Figure 9). They enable children to create their own robotic environments using LEGOs with gears, motors, sensors and *computation*. With the original LEGO bricks, children could build *structures*. With the addition of motors, gears, and other sophisticated mechanical parts, children could then build *mechanisms*. With the addition of computation, children can now build *behaviors*.

There are two approaches to building these behaviors: The first involves the new LEGO/Logo bricks that incorporate simple electronic devices such as sensors (light, sound, touch, and proximity), logic (and-gates, flip-flops, timers), and action bricks (motors, lights). These bricks can be connected to create simple behaviors. For example, hooking a sound-sensor brick to the motor drive of a car through a flip-flop brick results in a car with the behavior of starting and stopping at any loud sound.

Another approach uses Electronic Bricks, whereby a very small programmable computer is built into the LEGO brick. This allows the creation of more complicated behaviors without the need for any connection to a desktop computer. These bricks are about the size of a deck of cards and incorporate a Motorola 68332 processor with 25K of nonvolatile RAM and a two-line LCD display. They also contain four programmable motors or light connections, can receive eight inputs (8 bits each), and have a microphone and speaker for sampling and creating sounds (at 20 KHz). Finally, each have six pairs of infrared transmitters and receivers for communicating with other bricks.

The LEGO/Logo programming environment of today is used in over 7,000 elementary and middle schools throughout the country—its technological sophistication and interface simplicity is a powerful combination. LEGO/Logo has thus proven to be a creative and educational environment for children.

***Immersive rooms—Allison Druin, Ken Perlin, et al.,[3] NYU Media Research Lab.*** Immersive environments bring such disciplines as theater, robot-

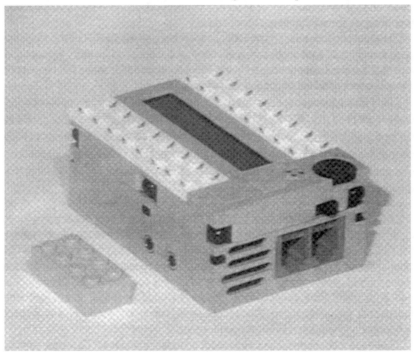

Figure 9: Mitch Resnick's LEGO/Logo project.

---

[3]The members of the interdisciplinary research team at NYU that have contributed their talent and energies to this work are as follows: Richard Wallace, Vineel Shah, Troy Downing, Raj Raichoudhury, Sabrina Liao, Jeff Severtson, Athomas Goldberg, Steve Bodo, and Kim Markegaard.

ics, film, environmental design, and multimedia together and enable a user to become an active physical participant in what can be an educational or entertaining experience. Lifting an object, touching a wall, or walking in a particular direction can make lights change and video images appear or create sound or voices. Sensor technologies are embedded into a physical setting, and when triggered can activate output that contributes to a person's experience in the room (Davenport, 1992; Druin & Perlin, 1994).

Today, researchers in universities and industry are developing immersive environments—physical spaces that are artistic statements, educational adventures, or entertaining experiences now found in museums, amusement parks, and even some automated homes. Researchers at the New York University Media Research Laboratory have been exploring the numerous possibilities of differing content, technologies, and design methodologies. Led by Allison Druin and Ken Perlin, this group's focus has been to develop the technologies and methodologies that nontechnical designers can use to create their own immersive environments.

Beginning in the Spring of 1993, 17 NYU Film students and faculty collaborated with 9 NYU Computer Science students and faculty. Out of this collaboration grew a visual programming application that communicated to an I/O controller. Using HyperCard on an Apple Macintosh Quadra, "a virtual patchbay" interface was developed that enabled immersive environment designers to associate any input (e.g., a tapeswitch, floor sensor, momentary contact switch) with any output (e.g., lights, sound, video). The I/O controller that was built could accommodate 16 standard AC-output devices as well as 24 digital-input switches.

This collaboration has also resulted in the development of various immersive environments for the public. One immersive experience was a light-hearted piece about baby-sitting; another explored the darker side of heaven and hell; the third was a suspenseful murder mystery.

Currently, the NYU Media Research Laboratory, in collaboration with the NYU Music Technology Group, is in the midst of developing an immersive musical environment (see Figure 10). In this experience, participants come to better understand the parts of music by interacting with physical objects in a room that represent such things as volume control (walking up stairs), different voices in the music (uncovering masks), and music texture or timbre (touching different physical textures).

## SUPPORTING TECHNOLOGIES

Building computer-augmented environments requires the use of various technologies, some of them standard and some of them experimental. They may be categorized into three classes: input, output, and combined input/output. The section that follows describes many of the technologies used to create computer-augmented environments.

## Input

Although traditional computer systems use keyboards and mice, computer-augmented environments rarely do. The technology of computer-augmented environments is integrated into the user's physical surroundings to form a seamless or sometimes even invisible interface. To this end, various input devices have been developed that attempt to give users intuitive ways of interacting with the computers behind the scenes.

Spatial locating devices are perhaps the most ubiquitous type of input system for computer-augmented environments. Usually attached to some part of the body, they inform the computer where the user, or a part of the user, is physically located. For example, in Feiner's system (Feiner et al., 1993) electromagnetic and ultrasonic sensors are used to track the orientation and position of the user's head, enabling the system to register computer graphics on top of the real world.

There are several technologies that can detect position and/or orientation

Figure 10: Immersive Room: Allison Druin, Ken Perlin.

which are summarized in Table 1. A majority of these sensors have two components: a reference part and the part that is attached to the person. Sometimes the
reference is the transmitter and the attached part is the receiver, and sometimes
vice versa. There are various tradeoffs between the technologies. Some of the
parameters that vary are range (maximum distance between transmitter and
receiver), accuracy, and working locations (indoors/outdoors). In addition, some
systems need a direct line of sight between transmitter and receiver.

Perhaps the best known system for spatial location is the *Federal Global
Positioning System* (GPS). Originally developed to help soldiers orient themselves in remote places, it has now been adopted for use by many industries.
Based on triangulation from a web of satellites (determining the delay between
signals from three or more satellites), a device about the size of your hand can
tell where you are anywhere in the world (location and altitude) within a few
tens of meters. Because a direct line of sight must be maintained with at least
three satellites, GPS cannot be used indoors, and may have trouble in some
cities with dense, tall buildings. Nevertheless, it seems well-suited for the trans-
portation industry, where it is being adopted to track trucking fleets and being
investigated to guide airplanes. GPS is also very commonly used for navigation
on commercial and private boats.

GPS is also being used for what is perhaps the biggest commercial application of computer-augmented environments. Japanese auto makers are currently
selling navigational systems for cars with electronic maps that show both where
you are and the best path to where you are going. They add audio to tell the dri-
ver when a turn is coming up that they should take. This is similar in spirit to
Bederson's automated tour guide system described here earlier. U.S. auto makers have announced that they will have similar products available in 1995.

*Electromagnetic* and *ultrasonic* devices are probably the most commonly
used devices for spatial location by both virtual- and augmented-reality applications. Like GPS, they both work by triangulation, but the first uses electromagnetic radiation and the second uses inaudible sound. They have similar characteristics, but the main difference is that electromagnetic devices are subject to
disturbance by nearby electromagnetic radiation and large pieces of metal, and
the ultrasonic devices must maintain a direct line of sight between transmitter
and receiver. For this reason, the ultrasonic devices seem to be more popular,
because most computer augmented environments are full of computers, displays, speakers, etc., which may cause interference. The most popular manufacturers of these products are Ascension, Logitech and Polhemus (Ascension
Technology, 1994; Logitech, Inc., 1994; Polhemus Technology, 1994).

Local radio is yet another method that uses triangulation. Similar in concept
to GPS, it requires three radio transmitters that are set up in the environment
and a small receiver circuit that can be carried around to locate itself. Local
radio has the advantage over GPS that it can be used indoors and has higher precision, but has the disadvantage that its range is much smaller, as only very
weak radio signals may be broadcast legally without FCC approval.

Digitized video combined with automatic computer vision processing is beginning to be used for tracking human participants. Although computer vision systems are difficult to create and may need expensive computers to run on, they can provide the most detailed information. For example, in addition to identifying where the participants are, they may also be able to identify *who* the participants are by recognizing their faces.

For some applications, it may be enough to know just the orientation of part of the user, and for these, *gyroscopic* systems are quite good. For instance, an application may need to know in just what direction someone's head is facing, or what orientation their hand is in without actually knowing where the person's head or hand is. Gyroscopic sensors have just recently become commercially available. They need only a single device (no transmitter/receiver pair), and are so small they can fit inside a standard 35mm film container. They are light enough to be mounted easily on a head-mounted display or on the back of a glove.

For some applications, it is necessary only to distinguish between several distinct places where the user is—rather than having a continuous description of user location. For these applications, it may be more appropriate (and cheaper) to place a single low-tech sensor at each point that needs to be identified. The automated tour guide project of Bederson and the immersive environments of Druin et al. fall into this category. For the former, the computer that is carried around by the user must know where the user is, and for the latter, the computer in the room must know where the user is. Because of this distinction in application, two different technologies are more appropriate for each.

The infrared system used by the automated tour guide project is custom-built, and works by placing infrared transmitters in the ceiling. Each transmitter projects a small cone downward and an ID unique to each location is transmitted once a second. Thus, a receiver can distinguish which transmitter they are underneath.

The switches used in immersive environments are placed at each relevant location. Many different kinds of switches can be used depending on what is to be detected. A person can be detected to be standing at a specific location by placing a switch under a mat. A person can be detected to be walking over a threshold by infrared transmitter/receiver pairs. A person can also be detected when objects are touched with microswitches or magnetic switches.

For some applications, it is necessary to have even finer information about the user than position and orientation. For a user to control the world with the gesture of his or her hands, the computer must have some way to find the exact position of the hand and fingers. There are two devices that are typically used for this: the DataGlove™ (made by VPL, now out of business; Zimmerman & Lanier, 1987) and the Exoskeleton™. The DataGlove is a fabric glove with thin fiber-optic cables along the back that detect how much each finger is bent by how much the light that is transmitted through the cables is degraded. The Exoskeleton is an elaborate mechanical device that attaches to the back of the hand with rings that go around each finger joint. It allows full motion of the hand and measures the bending of each joint with miniature potentiometers.

## Output

Superimposing images on the physical world is generally accomplished with output devices that are carried by the user. These images are typically generated by a special device that is worn on a person's head. It contains a half-silvered mirror placed in front of one eye that reflects the image from a tiny monitor mounted just above the mirror. One popular device, the Private Eye™ (Reflection Technology, 1994) is a very small 320 x 200 pixel monochrome display that can be adapted for use in this way (Feiner et al., 1993).

Table 1: Spatial Locating Devices

| Technology | Location | Range | Accuracy | Comments |
|---|---|---|---|---|
| Global Position-ing System (GPS) | Outdoors | No Limit | 10 s of meters | Line of sight, Position only |
| Electromagnetic | In/Outdoors | < 5 Feet | 1/100 s of Inches | Electromagnetic interference |
| Ultrasonic | In/Outdoors | <5 Feet 100° | 1/100 s of Inches | Line of sight |
| Local Radio | In/Outdoors | 100 s of feet | Inches | Position only |
| Computer Vision | In/Outdoors | ~90° | Inches | Dependent on implementation |
| Gyroscope | In/Outdoors | 360° | Degrees | Rotation Only |
| Infrared | Indoors | < 20 Feet | Feet | Line of sight, Position only, Inexpensive |
| Switches | In/Outdoors | No Limit | Exact | One switch per location, Inexpensive |

Superimposing audio is much easier and simply requires standard head-phones. However, a portable general-purpose computer-controlled random access digital audio storage device does not yet exist commercially. One researcher (Bederson et al., 1994) addressed this problem by opening up a Sony MiniDisc™ and attaching wires to each button. The player could then be controlled externally by applying voltages to the wires to simulate pressing buttons.

Immersive environments require a different set of output devices as they imbed computer-generated data directly in the environment. Here one would typically find projected video displays, slides, movies, etc. In addition, atmospheric changes in sound and lighting can be computer-controlled. Motors of various kinds (rotary, linear, spherical) can also be used to physically move objects in the environment (Wallace, 1993).

## Input/Output

Some of the more novel devices combine both input and output in a single unit. Force feedback or haptic devices measure human motion and use a motor for output to provide direct physical feedback to the user. Force feedback joysticks, for example, have been used to give a realistic sensation of using a motorized cutting tool (Balakrishnan, Ware, & Smith, 1994) and of moving over complex textures (Minsky, 1990). Haptic interfaces are not yet commonly used in computer-augmented environments.

On the other hand, devices that integrate pen-based input and screen-based output are becoming quite common. In the past year, Personal Digital Assistants (PDAs) have come on the market. A PDA is a very small portable device that has an LCD display with a digitizing tablet directly under the display so that the user can electronically write directly on the display. One group of researchers have been exploring this concept for several years and have built several prototype devices in three different scales (see section on Weiser, this chapter).

## Future Technologies

What the future holds is, of course, impossible to know. However, based on the current directions in which researchers are heading, we can make some assumptions on where these hardware and software technologies are moving.

One active area of current research that will likely yield more practical results in the near future is the recognition of natural human movement and speech. Many researchers are currently working in this diverse field (Baudel, 1993; Bolt, 1980, 1984; Brooks, personal communication, April 1994; Koons, Sparrell, & Thorisson, 1993; Schmandt, 1992). Using *free-hand* gestures may someday be a natural, terse, and powerful method for interacting with augmented computer environments. Since Krueger (1991, 1993) first started exploring this form of interactivity in the 1970s, there has been quite a lot of work in this area. However, to date, there still have not been any commercially successful applications using this technology. There are many reasons for this; as discussed in Baudel (1993), the following list reveals many problems that must be faced:

o *Fatigue*: Using free-hand gestures is tiring. Moving a mouse and typing uses much less energy than holding the arm up and moving the whole hand. Gesture interfaces should be as natural and fast as possible with a minimum of movement.

o *Learning Curve*: It may not be apparent to the user of a gesture-based system what the available gestures are, and what they do. Gesture interfaces should be easy to learn and provide immediate feedback.

o *Lack of Comfort*: Most gesture-interface systems today require that the user wear some type of glove that attaches to the computer. This restricts mobility and makes such systems unnatural. However, the early work of Krueger (1991, 1993) and of Wellner (1993) are examples of computer-vision-based gesture interfaces.

o *Gesturing versus Natural Hand Movement*: It is difficult for the system to distinguish when a gesture is intended to affect the system, and when a gesture is simply a natural human movement. Unless the system can always distinguish between these, there must be some very simple mechanism for the user to specify when a gesture is intended for the system.

o *Segmentation of Hand Gestures*: As with speech recognition, it is very difficult to understand the continuous motion of the user's hand in terms of discrete actions. For this reason, many existing systems operate only on static gestures. Work on dynamic gestures is progressing as discussed in the next section.

One research group is looking at gesture recognition specifically to address these issues. Baudel (1993) and Beaudouin-Lafon developed the Charade free-hand gesture system to address the aforementioned problems. Charade lets the user's hand gestures control a computer-aided presentation. The user wears a VPL DataGlove connected to a Macintosh computer with a projected display. The user specifies when a gesture is to be recognized by pointing to the projected display. For each gesture, the hand moves from a start position to an end position with different finger positions at start and finish. Gestures always start with a tense position and finish with a relaxed position to try and alleviate the aforementioned fatigue problem. There are 17 gestures, such as "next page," "previous page," "next chapter," etc. Baudell and Beaudouin-Lafon reported that first-time users had a 72 to 84% gesture recognition rate, whereas trained users had a 90 to 98% success.

The use of speech as a general interface to computers has been a long-soughtafter goal. As computers get more powerful, this possibility becomes more and more likely. Although the keyboard may never be completely replaced, there are many applications in which voice recognition will revolutionize the way we interact with computers. The use of speech recognition in computer-augmented environments is easy to guess. Much of the focus of work in this field has been in making the interface to technology simpler and more natural—and what could be more natural than talking? (See Schmandt, 1992, for a review of voice in computing.)

Another area of work (which will be important for more practical augmented video displays) is in small, light, very high-resolution displays. A group at the MIT Artificial Intelligence Lab (Brooks, personal communication, April 1994) is developing novel microdisplays by developing techniques for making displays out of standard integrated circuit technology. Using CMOS circuitry, they can use off-the-shelf fabrication technology, which will yield denser and cheaper circuits as the standard technology improves. By creating a chip with five micron square pixels, they will be able to construct a 2,000 by 2,000 (4 megapixel) display in a one square centimeter active area. Such a device will use very little power. Current LCD displays require most of their energy for lighting the screen. Because these microdisplays will be mounted very close to the eye, they will need very little light, and thus much less power.

We expect that haptic interfaces will continue to be an area of active interest. Although bitmap displays are today's standard for information presentation, humans have an entire body full of sensors and a lifetime of experience in interpreting physical sensations. As the development of haptic interfaces continues, we will be able to tap this rich modality of interaction. Massie (1993) recently developed the PHANTOM haptic interface, which couples directly to the fingertip or to a stylus that the user holds. This device can generate forces along three axes in response to fingertip motions. Massie's team has been able to generate forces resulting in sensations of contact with rigid and compliant objects, texture, friction, object impedance, and motion. They have also begun to construct demonstrations of more complicated procedures including virtual button pushing, epidural needle insertion, and a prostate examination.

### Future Applications

In the future, computer-augmented environments may be found in every place in our lives: in our homes, work spaces, educational institutions, and places of entertainment. Any physical environment may in the future be enhanced by superimposing computer-generated information on it, and/or embedding computational technologies into it.

Today researchers are experimenting with various technologies to create targeted applications for specific spaces (e.g., museums, schools, work environments). Tomorrow we expect these numerous technologies to be integrated into all parts of many different familiar environments. The same technologies can support many different activities.

These technologies will actively enhance our physical surroundings. To do this, computer-augmented environments will have multimodal interfaces. They will recognize our gestures, understand our speech, respond to our touch, and interpret our drawings and text. In response to these multiple forms of input, computer augmented environments may respond by verbally discussing a situation with a person in the room. They may superimpose information that seems useful for the situation at hand and may move objects in the room for more efficient activities. Multimodal input and output can create very powerful possibilities for computer-augmented environments.

One new large project that is just beginning aims to take advantage of this work by creating an immersive environment in what they call the *Intelligent Room*. Directed by Rodney Brooks at the MIT AI Laboratory, this research envisions a room that will seamlessly interact with participants, effectively eliminating an explicit interface to sophisticated technology. The participants will control the room with their natural gestures and speech. Using computer vision, the room will recognize and track multiple occupants and recognize hand gestures. The use of *enhanced reality* glasses will add registered computer-generated graphical data to the room. Voice recognition will also be used to access information and to control the room. Finally, there will be software agents that will provide access to services on the National Information Infrastructure. Agents will attempt to understand the high-level goals of the participants and will automatically attempt to perform some of the simpler chores necessary in attaining them.

In the scenario that follows, our vision for the future is described in further detail. It may sound like science fiction, but we believe it can soon be a reality. By integrating many diverse pieces of existing and emerging technologies, computer-augmented environments can automate, facilitate, and collaborate in our future activities.

## ONE DAY IN THE FUTURE

It is morning. Danielle Brown is just waking up. She opens her bedroom door to hear a message from her mom. "Danielle, I had an emergency at work, and had to leave early. It looks like a part of the building we just renovated is weak. I'll tell you the details later. I left a message with Mr. Lane asking him to drive you to school, but you may want to call him when you get up to confirm. Be good! See you later."

Danielle wanders into the kitchen to grab a glass of juice. As she's drinking, she says, "Phone, call Mr. Lane" and quickly checks to see if she's wearing her pajamas with "smart stripes" that automatically shut off the video option of the phone because she does not want Mr. Lane to see her in pajamas.

Danielle hears the voice of Mr. Lane say, "Hi, is that you Danielle?"

"Yes, Mr. Lane. My mom left me a message to call you this morning."

"That's right, I'll be picking you up in half an hour for school. Your mom wanted me to bring you by the building she's working on later this afternoon. The automatic navigation system has a kink in it, but your mom's already downloaded the directions to my car."

"Thanks for helping us out!"

"No, problem. See you soon."
Danielle finishes her last gulp of orange juice and runs to shower and dress. Just as she finishes her daily routine she hears Mr. Lane beeping his car horn outside.

She arrives at school and immediately begins working on her assignment. The members of Danielle's geography team have not yet all arrived, so the few that have begin by moving their assigned work table to the back corner of the room. When they do so, all of the chairs associated with that table move around the table themselves. They sit down and begin work.

"I can't believe we have to work with kids in Greece! I wanted to work on Switzerland this time," says Bob.

Jen says, "It could be worse, we might have had Russia like Sarah's group. They have the worst Internet connection of anybody. You can never hear them, let alone see them."

Just then the teacher comes by. She sits down at the work table and says, "Retrieve geography assignment 4." The work table video screen then shows them the maps and information they have to work with. At the same time their box of work sensors pops out of a compartment in the table surface.

"Now I just want to remind you of a few things," their teacher says. "This time you need to create an environment which gives me the feeling that I am in Greece. You can use any of the team sensors in your box to make your environment, but just remember that your heat sensor doesn't seem to be working properly. So if you want some climate control in your environment you'll have to fake it with another sensor. Oh, yes, if you want to get in enough time with your partners in Greece, you better get to it. Remember your time difference—they'll be going home soon. Any questions before I go on to the Russian Team?"

"Yes. I was wondering," said Bob, "are we supposed to pick a present-day place in Greece, or any one throughout history? I was reading yesterday about Greece during World War II, and there was a lot of interesting stuff going on then." As he says this, a copy of the paper he was reading comes upon their video table top. It begins showing a video clip.

The teacher replies, "If you can tell me everything I'd want to know about Greece, then sure, try it. It may be harder, though. Remember that I want you to incorporate at least 50% physical models and 50% virtual information. It's always more time consuming to embed sensors in models than it is to just work with the video wall, but I think you can do it!"

The team groans, and the teacher moves on.

After school, Danielle is driven by Mr. Lane to her mom's building. They listen to her mom's directions as Mr. Lane drives. "Turn north at the next block. The building you are looking for is on the left side of the street, number 25."

When Danielle arrives she finds her mom outside the building in a conference with her partner at a remote site. Her mom is wearing annotating glasses that show the building's blueprints and is looking at the building. She is deep in conversation with her partner. "But Will, from what I can see on the blue prints the support structures should not have fallen in here. I've gone through the building five times today. I go into the living room and I can hear all of our design notes. According to what we did in those design sessions, it seems we should be just fine, but we're not. When I go into the kitchen, the notes about what we did there confirm that. Oh, wait Will, my daughter just arrived from school. Can I get back to you? Thanks!"

They greet each other, and fill each other in on the day's events. When they walk into the trailer office they see the wall calendar blinking. Danielle's mom touches the calendar and they hear, "You have tickets for a basketball game. Would you like to keep your tickets?" Danielle's mom says, "Yes, please."

Before they leave, Danielle's mom leaves some notes for her partner on the whiteboard wall in the trailer. She will retrieve them back at the site later tonight. As they close the door of her trailer office, they hear a voice that says, "You seem tired, should I shut down everything for the night?" Danielle's mom answers, "I wish you could, but unfortunately I'll be back to work later. Thanks." They shut the door and leave for their game.

While driving to the game, Danielle is describing her class project when her mother interrupts her. "Danielle, I just figured out what happened with that building today. Hold that thought." She slides out the tablet from the car console and says "Car, please autodrive for a few minutes" and then to Danielle

"Keep an eye on this thing, I still don't quite trust it." She sketches a picture of the building as she remembers it on the tablet with her finger. She intentionally leaves out an important beam that she now realizes the contractors somehow left out. As she sketches, the image appears on the flat panel display ordinarily used for displaying maps. She says "Car, please fax this to Will and I'll take over driving now. Thanks."

## REFERENCES

Ascension Technology. (1994). Ascension Bird product literature. Burlington, VT.

Balakrishnan, R., Ware, C., & Smith, T. (1994). *Virtual hand tool with force feedback.* Companion of SIGCHI `94 conference, pp. 83-84.

Baudel, T. (1993). Charade: Remote control of objects using free-hand gestures. *Communications of the ACM, 36* (7), 28-35.

Bederson, B. B., Davenport, G., Druin, A., & Maes, P. (1994, ). *Immersive environments: A physical approach to multimedia experience.* Panel at IEEE International Conference on Multimedia Computing and Systems.

Bolt, R. A. (1980). Put-that-there: Voice and gesture at the graphics interface. *Proceedings of ACM SIGGRAPH, 14*(3), 262-270.

Bolt, R. A. (1984). *The human interface.* Belmont, CA: Wadsworth.

Caudell, T. P., & Muzzle, D. W. (1992). Augmented reality: An application of heads-up display technology to manual manufacturing processes. In *Proceedings of the Hawaii International Conference on System Sciences* (Vol. 2, pp. 659-669).

(1993). *Communications of the ACM* [Special Issue on Augmented Environments], *36* (7).

Davenport, G., et. al., (1992, December). *Presentation of immersive environments research,* MIT Media Laboratory, Cambridge, MA.

Drascic, D. (1993). Stereoscopic vision and augmented reality. *Scientific Computing & Automation, 9* (7), 31-34.

Druin, A. (1988). Noobie: The animal design playstation. *Proceedings of ACM SIGCHI, 20*(1), 45-53.

Druin, A., & Perlin, K. (1994). Immersive environments: A physical approach to the computer interface. *Proceedings of SIGCHI.*

Druin, A., & Perlin, K. (1995). *Immersive music room.* New York: New York University Media Research Laboratory.

Feiner, S., MacIntyre, B., & Seligmann, B. (1983). Knowledge-based augmented reality. *Communications of the ACM, 36*(7), 52-62.

Fitzmaurice, G. W. (1993). Situated information spaces and spatially aware palmtop computers. *Communications of the ACM, 36* (7), 38-49.

Furness, T. (1969, May). *Helmet-mounted displays and their aerospace applications.* Paper presented at the National Aerospace Electronics Conference, Dayton, OH.

Gold, R., Buxton, B., Feiner, S., Schmandt, C., Wellner, P., & Weisner, M. (1993). Ubiquitous computing and augmented reality. *ACM SIGGRAPH Panel in Computer Graphics Proceedings, Annual Conference Series* (pp. 393-394).

Koons, D. B., Sparrell, C. J., & Thorisson, K. R. (1993). Integrating simultaneous input from speech, gaze and hand gestures. In M. T. Maybury, (Ed.), *Intelligent multimedia interfaces* . Cambridge, MA: AAAI/MIT Press.

Krueger, M. W. (1991) *Artifical reality II*. Reading, MA: Addison-Wesley.

Krueger, M. W. (1993). Environmental technology: Making the real world virtual. *Communications of the ACM, 36*(7), 36-37.

Laurel, B. (1991). *Computers as theater*. Reading, MA: Addison-Wesley.

Laurel, B. (Ed.). *The art of computer interface design*. Reading, MA: Addison-Wesley.

Logitech, Inc. (1994). Logitech 6D mouse product literature. Fremont, CA.

Massie, T. (1993). *Design of a three degree of freedom force reflecting haptic interface*. Unpublished bachelor's thesis, MIT, Cambridge, MA.

Minsky, M. (1990). Feeling and seeing: Issues in force display. *Computer Graphics, 24*(2), 235-243.

Papert, S. (1980). *Mindstorms: Children, computers and powerful ideas*. New York: Basic Books.

Polhemus Technology. (1994). Product literature.

Reflection Technology. (1994). Private Eye product literature. Waltham, MA.

Resnick, M. (1993). Behavior construction kits.*Communications of the ACM, 36*(7), 64-71.

Schmandt, C. From desk top audio to mobile access: Opportunities for voice in computing. In H. R. Hartson & D. Hix (Eds.), *Advances in human–computer interaction* (Vol. 4, pp.251–283). Norwood, NJ: Ablex.

Seligmann, D., & Feiner, S. (1991). Automated generation of intent-based 3D illustrations. *Proceedings ACM SIGGRAPH '91, Computer Graphics, 25*(4), 123–132.

Streeter, L. A. (1985). How to tell people where to go:  Comparing navigational aids. *International Journal of Man–Machine Studies*, 549–562.

Sutherland, I. E. (1968). A head-mounted three dimensional display. *Fall Joint Computer Conference, AFIPS Conference Proceedings, 33*, 757–764.

Wallace, R. S. (1993). Miniature direct drive rotary actuators. *Robotics and Autonomous Systems, 11*, 129–133.

Want, R., Hopper, A., Falcao, V., & Gibbons, J. (1992). The active badge location system. *ACM Transactions on Information Systems, 10*(1), 91–102.

Weiser, M. (1993). Some computer science issues in ubiquitous computing. *Communications of the ACM, 36*(7), 74–84.

Wellnar, P. (1993). Interacting with paper on the DigitalDesk. *Communications of the ACM, 36*(7), 86–96.

Zimmerman, T., & Lanier, J. (1987). A hand gesture interface device. In *Proceedings of human factors in computing systems* (CHI '87) (pp. 235–240). New York: ACM Press.

# Chapter 3
# Agents: What (or Who) are They?

Katherine Isbister
*Communication Department*
*Stanford University*


Terre Layton
*SunSoft*
*Sun Microsystems, Inc.*

## WHAT ARE AGENTS?

"Excuse me, but your mother is on the phone. She wants to remind you to pick up the birthday cake."

[Able (robot) open(door)5]

"Brochure wizard..."

Each of the above is an artifact of an "agent." The first example is from the Apple Knowledge Navigator video's "Phil" agent; the second is a statement of an agent's capabilities in Shoam's (1990) article on agent-oriented program- ming; the third is a pull-down menu choice in Microsoft Office Powerpoint™. How can such disparate things be called agents? All three involve computers emulating human qualities in some way; in particular, human abilities to take actions and to communicate.

For a long time, the prevailing metaphor for computers was "tool": the computer is a tool for humans to use, to customize to do efficient work; a tool for communication and for writing and editing; for calculating figures and keep- ing track of complex information. All of these things are still true about com- puters, but there are uses that aren't well described by the "tool" metaphor.

People now use computers to communicate with other people, or other computers. Those who learned about computers by using isolated desktop units can work on server/client systems or connect to large networks of machines through the phone lines. In addition, more powerful computers are able to perform many tasks at once. A user may have two or three tasks running in the background, and be in the middle of another complex process. How do we explain and understand an object that can communicate with others like it, patch us through to other people, go off and do several things at once, remind us about ongoing projects and appointments? Such an object seems inherently different than a hammer or wrench—it has active qualities. It acts on one's behalf—it is an agent.

The idea that a computer could act as an agent is not new. According to Kay (1984), it:

> originated with John McCarthy in the mid-1950s, and the term was coined by Oliver G. Selfridge a few years later. . . . They had in view a system that, when given a goal, could carry out the details of the appropriate computer operations and could ask for and receive advice, offered in human terms, when it was stuck. An agent would be a "soft robot" living and doing its business within the computer's world. (p. 58)

Over the past 10–15 years, advances in programming techniques and computing power have made agents an increasingly plausible idea. Human–computer interface specialists have declared that the future of the interface will include agents—Negroponte, quoted in Laurel (1990), said: "In some form we can expect surrogates who can execute complex functions, filter information, and intercommunicate in our interest(s)" (p. 352). They cite the natural tendency of people to anthropomorphize:

> Interface agents draw their strength from the naturalness of the living-organism metaphor in terms of both cognitive accessibility and communication style. Their usefulness can range from managing mundane tasks like scheduling, to handling customized information searches that combine both filtering and the production (or retrieval) of alternative representations, to providing companionship, advice, and help throughout the spectrum of known and yet-to-be-invented interactive contexts. (p. 356)

This chapter focuses on the human interface aspect of agents. As Kay (1984) stated, "what is presented to one's senses is one's computer" (p. 54) Although there is a significant body of work on intelligent agentry in the artificial intelligence field, much of it focuses on interagent communication techniques. These may be the backbone for powerful intercomputer communication for users, but it will be transparent to them. These techniques are thus not covered in this chapter, except insofar as they are linked to new functions presented to users with an agent approach. We instead focus on agents in the interface—links from the user to the machine, and propose a broad framework for describing and understanding agents, based on the human qualities they emulate. This does not mean that we advocate encouraging anthropomorphism. Rather, we are proposing that what is new about "agents" are the new realms of human activity that are being imitated. Just as a hammer extends the hand, an agent extends the social sense of a user. This does not mean that the person thinks the agent is a pers on, merely that the person is using his or her understanding of agency to predict the agent's actions and to communicate with it.

You won't find a detailed, task-based taxonomy of current agents in this chapter (Wood, 1993, offers a helpful task-based taxonomy in his paper available on the WorldWideWeb). We have included examples of current agents that illuminate the classes of agent "intelligence" we describe, and guidelines for how to apply what we found in our survey of current agent research. We hope this chapter will give you an overview of some of the interface issues that arise when designing agents, and provide a point of reference for further exploration.

## I STILL DON'T KNOW WHAT AGENTS ARE...

Knowing that interface agents emulate human action, assistance, and communication still doesn't give a clear picture of what exactly agents are. How do agents emulate human action? How do they assist users? How do they communicate? What do they look like? The answers vary widely. There is much debate over what should and shouldn't be called an agent. Our definition takes the broadest possible approach. It encompasses everything from dialogue-box-based assistance to learning agents. Rather than focus on what tasks or techniques define an agent, we've focused on the source of the agent concept—people. People act, assist, and communicate. An interface agent is something designed to emulate these human abilities. Different agents mimic different human attributes in different ways, depending on their purpose. In the following section, we've listed human qualities that agents emulate.

### Human Qualities That Agents Emulate

*Job description*. In the workplace, job description becomes a large part of who someone is. The same is true of agents in the interface—they can be thought of as entities with assigned duties like filtering mail, offering advice on programming, reminding you of appointments and obligations, and helping you draw geometric figures. This is slightly different from the tool metaphor, in that agents may go off and perform their duties on their own. This makes other attributes like control and trustworthiness important (see "Taking Control" next example).

*Conversational ability*. We have already leveraged our communication skills in computing—using commands to get the computer to do things is assigning it at least primitive status as something that can communicate. In some sense, the graphical user interface was a move away from conversation toward manipulation of tools. The agent concept can bring more conversation to the interface where it's appropriate, using human communication as a model (Laurel, 1990). Agents may have a wide range of conversational skills; some only understand choices you make in dialogue boxes,whereas others use a command line. Agents may also be created that comprehend and act on natural language, or even nonlinguistic gestures or cues.

Understanding is only half of conversation—it's also important that the agent can communicate back to the user. This is potentially a great strength of the agent concept—a greater focus on using existing human communication skills. An agent can give feedback like error messages in a way that the user understands, check in with the user to make sure all is well, ask for clarification if it can't do a task as requested, respond to a query and so on. However, an agent needn't mimic all human conversational abilities. It might be enough for an agent to give a limited repertoire of prewritten replies to certain queries. The point is to use human communication skills as a starting point for shaping the interaction style of a particular agent—using what we already expect in human conversation to create an intuitive interface (Clark & Brennan, 1991).

*Taking control.* One of the potential advantages of the agent concept is having someone or something to which to delegate tasks. The typical graphical user interface presents all the options available, giving the user total control over a carefully limited range of abilities. An agent could act as an intermediary or gatekeeper to processes on which a user might not be an expert. It is often nice to relinquish control over some aspects of one's environment—most people don't want to have to fly the plane they take to another city, or design the highways on which they drive (Gentner & Nielsen, 1994).

Delegation means handing some control over to the person or machine to whom you've given responsibility for a task. Among people, negotiating this handoff is a tricky and ongoing process. An agent can be designed with a range of control, from volunteering advice only on request to making high-level judgments about what sort of information to which the user will have access (Foner, 1993). Users may be able to adjust how much control the agent has, depending on their personal preferences. One user might only want to directly assign tasks of the moment to the agent. Another user might feel comfortable assigning tasks to be done later, or routine tasks to perform when needed.

Because the computer has the ability to monitor all the user's activities, control issues extend to what sorts of uses the computer can or should make of this sort of information. It's possible for the agent to watch the user's actions, and then try to predict behaviors, and automate them if the user wishes (Maes & Kozierok, 1993).

*Trustworthiness.* Trustworthiness is a quality that creeps in with agency— in general, graphic interfaces imitate tools that do not take responsibility for their actions—any error is an error on the user's part in understanding the limitations of the tool in hand. Agents allow the user to turn over control, but force decisions about when to do so onto both the interface designer and the user. Just as among humans, the amount of control a user will be willing to give to an agent will depend on how trustworthy the agent seems (and on how risky the action is) (Foner, 1993). A user may initially be reluctant to allow the agent to automate any of his or her actions, but over time, as the user sees the agent predict actions accurately, may be willing to turn over some control (Maes, Darrell, Blumberg, & Pentland, 1994; Maes & Kozierok, 1993). A well-designed agent could build up trust slowly and steadily through its actions, and be designed not to overstep its abilities.

*Looks*. In general, human beings share a set of standard physical characteristics, as well as a tendency to adorn/ornament themselves with things that can convey a wealth of information about status, tastes, interests. Humans change appearance with age and with the types of things they choose to do that end up altering how they look. Agents can be designed to take advantage of the elaborate expressive codes we humans are constantly creating and altering (Laurel, 1990). A raised eyebrow could be an effective "Are you sure you want to do that?" A pilot costume on the agent could indicate that it is a navigation expert. An aged face could indicate wisdom and experience. Using the agent concept opens a wealth of nonlinguistic communication possibilities for the interface. There are already examples of use of these sorts of codes in iconography and illustration in the graphic design world, as well as in film, movies, and the arts in general. Agents can take advantage of the systems of meaning and expression developed in the arts around the human form.

*Personality*. Even though the jobs that people fill are structured in a certain way, each person brings something unique to the role. Personality is used here as a catch-all term for human features and behaviors and the individual style they represent. It involves the appearance of the agent—How is it drawn? What sorts of features does it have? Animators have honed the communication of personality through illustrative style. Imagine being able to use all the cues animators use through appearance and behavior to show what an agent can do and give a hint as to how it will approach tasks. An agent may also have a distinctive interaction style—it might be shy and infrequently speak up, or might be very confident, offering friendly and firm suggestions to a hesitant learner. This is a close adjunct to looks—it's the overall impression of the individual that emerges over time through interaction. If one wants to create an agent that will be a constant desktop companion, it will be important to think through how its personality will emerge over time. People tend to change their behaviors and learn new things which get incorporated into the routine, often picking up mostly things that conform to previous personality. Believable agents may require similar shifts in order to remain the same over time.

## DEVELOPMENTAL STAGES OF AGENTS...

When reading through the literature on agents, one thing became very clear early on—there is no common language used to describe the various types and/or classes of agents. Often you find that people appear to be talking about different types of agents, when really they are referring to the same thing, it is just the lack of a common vocabulary that separates the two (and the reverse is also true). One of the goals of this chapter is to provide some dimensions of agency that researchers and practitioners alike can reference when referring to the use of agents in the interface.

Several people have attempted to classify agents; some have based their descriptions on the agents' task and one we found based their agent descriptions on business processes. Both of these approaches we found fell short in terms of providing clear descriptions that had boundaries with no overlap. It is very difficult to define agency in terms of tasks or goals because every task is made up of multiple subtasks that may or may not overlap with some other task or goal. If the goal of defining some dimensions for agency is to give researchers and practitioners a common vocabulary that is clear, it is much better to come up with dimensions that have clear boundaries. One thing that appears to work better is to think of the dimensions of agency as we think of human development, beginning from the time someone is an infant.

One common reaction to interface agents that have been developed is to find fault with them for not emulating human behavior realistically enough. Rather than labeling anything that falls short of a fully developed adult human intelligence as a failure, we think there is a continuum of agent development, not unlike the stages that Piaget used to describe his observations of children's development. Moreover, human intelligence is a dubious goal for software, and one that has proven intractable. We do not believe agents should (or ever could) be human—agents use humanness as a metaphoric guide.

Some of the things that we feel are important when defining these dimensions of agency are listed later. As stated previously, to describe the dimensions of agency we felt that it was best to represent agency as existing along some continuum, where the stages are considered to be serial steps. At each stage along this continuum, each stage subsumes and integrates the abilities of the previous stage, and the development along this continuum is cumulative (i.e., at each stage, a restructuring takes place with the result being something new that is more knowledgeable/competent). Also, there is never a regression in agent abilities along this continuum.

This section presents a possible solution to the agent-definition problem that we have based on Piaget's Theory of Development. The reader should not focus so much on the actual theory and/or the validity of the theory, but more on the similarity and comparison of how agents develop. We use Piaget's theory as a metaphoric guide for our dimensions of agency, which appears to work very well.

At its most basic level, an agent will have specific abilities that it can perform for a user, but be sophisticated enough to be autonomous and can make decisions on its own and do tasks without needing instructions (this also includes agent to agent communication). This approach ties Piaget's theory in with our defined dimensions of agency to help describe agency existing along some continuum, whereby the agent becomes more and more knowledgeable and adaptive (as it moves along this continuum) to the users' environment. We've grouped examples of current agents in the following sections along this developmental dimension, where agents at each developmental stage may be quite effective, despite their obvious lack of total human intelligence. (It's important to separate the development of agency outlined later from the attributes we have previously listed. Those attributes are part and parcel of a developing human being, but as Piaget did, we are focusing here primarily on the agent's "intellectual" development—how it thinks, or appears to think.)

*Mere personification (just a gleam in someone's eye...).* The first stage doesn't really correspond to Piaget's scheme—it is the stage in which the agent concept works purely on the iconic level. Mere personifications do not really offer any intelligence to the user. They are communication devices that act as bridge objects in the interface—to help show the user what options they have, where they can go. These agents could be compared to document and application icons—they fit within the GUI environment, and offer added explanatory ability and continuity from place to place.

*Examples.* Microsoft Fine Artist and Microsoft Creative Writer agents are two programs geared toward children aged 8–14. Each has an agent character that appears as a cartoon child. The characters don't move or speak, with the exception of a few animated hops across the screen when illustrating a point. Instead, they work as adjuncts to hypercard-like stacks of information. On each "card", the agent might offer advice through a popped-up balloon of conversationally worded text. The user can click on the agent to be guided through tutorials on drawing tricks, greeting card construction, and other topics. The agents are part of a playhouse metaphor for the interface. They are cast as playmates who can help the user get started on a project.

There is also an agent common to both programs, called "McZee." This agent is not a playmate, but rather a sort of magical creature that helps the user navigate around the playhouses, and explains the use of drawing and writing tools. He, too, is a clickable object that has conversational text advice which can be stepped through.

These agents function more as organizing icons than as actors. They give the users a place to go for information that might most logically come from another person who knows the ropes. In this case, social expectations can be used to make the program less threatening, cold, or obtuse, without adding sophisticated dialogue capabilities.

The Apple Guides project, described in Laurel's (1990) *The Art of Human-Computer Interface Design*, is another example of the more iconic approach (not to be confused with Apple's new help system). The Guides project team tried to design guides that would help users of an educational hypermedia database find information of interest. The guides were intended to provide users with a narrowed range of topics to view next, and each guide had "interests" that matched the iconic figure of the guide. For example, a pioneer woman guide might take the user to areas of interest to someone like herself. The designers created a life story for each guide, and tied items to the guide accordingly. The user can click on a depiction of the guide to choose among his or her suggestions.

This navigational model met with mixed success (Laurel, 1990). On the one hand, users seemed to enjoy the guides; however, they weren't always able to understand why a guide "picked" a topic. Some users took the organizational metaphor further than the designers intended—attributing emotions to the guides or thinking that the information they were getting might be biased based on what the guide's perspective on the events would be. The users seemed to want more anthropomorphic qualities, given the human metaphor. The design group came up with some interesting conclusions about the power of anthropomorphism and characterization (for a detailed account, see Laurel, 1990).

This project reveals some of the potential problems or challenges involved in using mere personifications. Creating a human-like figure on the screen that the user is encouraged to go to for assistance may lead to overblown expectations about the human-like capabilities of the "character." Personifications normally offer little feedback to the user, and do not engage in dialogue. It's probably best to use them in contexts where this is expected and appropriate, or made clear through other design choices.

*Naive agent (sensorimotor stage).* This sort of agent's intelligence is of a "What's in front of it" variety. It can perform tasks that it already knows how to do immediately on request. This type of agent generally responds only when direct request are made or specifically scheduled. It doesn't make generalizations or abstractions, and it isn't able to determine for itself when tasks ought to be done. However, it can be quite effective at performing commonly requested tasks that would be tedious or overly complicated for the user, and can be an effective "teach-by-showing" tutor.

*Examples.* Microsoft Wizards show users how to do certain tasks within various Microsoft windows programs. For example, a user can select the brochure agent from a menu in the Powerpoint program, and step through a series of dialogue boxes. Within each box, the user chooses an option concerning the desired finished product, and the agent uses these choices to generate a sample document for the user. This is a good example of an agent that can respond to direct input and use its knowledge to perform a task for the user. This sort of agent can be very useful as a tutor—seeing a task being stepped through is a powerful way to learn how to perform that task. (See chapter this volume, for more information on the rationale for task selection and the development of these agents.)

Apple has developed a help function—Apple Guide—in System 7.5 that can walk the user through the task she or he wishes to accomplish. The user chooses Apple Guide from the help menu, and can then choose a help topic and a subtopic (such as "How do I adjust the speaker volume?"). The guide can also search for a term. When the guide walks the user through the desired task, it acts as a coach by circling or highlighting the object onscreen that should be manipulated next. It can adjust if the user works ahead, and will show the proper step should the user make a mistake. The demonstration tasks have been programmed into the help system, and it is built so that additional tasks can be programmed in by developers or system administrators. This is another effective use of a naive agent to tutor by example.

*Knowledgeable agent (preoperational stage).* The majority of today's agents fall under this category. Their "intelligence" corresponds roughly to Piaget's preoperational stage in that these agents can usually classify and sort things for the user, and are often able to use language in more sophisticated ways than the naive agent. They support asynchronous assignments and may be able to learn from observing the user over time. Knowledgeable agents tend to be more proactive and anticipate users' needs.

Knowledgeable agents may also act as advisors, teaching about specific areas of expertise through contextual examples/suggestions adjusted to the user's use patterns and needs. They're able to observe and offer ongoing assistance that changes as one learns.

*Examples.* Hewlett-Packard's NewWave agents are part of an object-oriented desktop that runs on top of Microsoft Windows. These agents are assigned tasks to perform through three modes: (a) users launch the Agent Task manually (e.g., dropping an Agent Task icon on the Agent Tool Icon); (b) Tasks are scheduled to be executed on a specific date, at a specific time; and (c) users set up events that the Agent monitors and executes when the conditions are met. Users create these Agent Tasks in two ways: (a) by having their actions recorded by the Agent and then having the Agent play the task back by using one of the three aforementioned methods; or (b) users write an Agent Task using the Agent Task Language—a simple scripting-type language designed to mimic normal English. Agents are designed to work throughout the environment, within and across applications. NewWave Agents provided a powerful facility for end users and developers alike. Both sets of users could off load tedious tasks and use their computers, even when they were away from the office (e.g., having the Agent do tasks such as backing up their machine while they are away or emptying the trash, etc.).

Researchers at MIT's Media Lab have created a series of agents that learn by observation of users, and apply their learned knowledge to various information filtering tasks for the user: mail sorting, calendar management, news group filtering, and music recommendations. In order to adapt to a user, the agent "gradually learns how to better assist the user by: observing and imitating the user, receiving positive and negative feedback from the user, receiving explicit instructions from the user, asking other agents for advice" (Edmonds, Candy, Jones, & Soufi, 1994, p. 40).

For example, the mail agent monitors the user's patterns of mail use—when does she or he delete a message, when is it saved? The agent looks for correlations between information, such as who sent the message or the date it was sent, and the subsequent behavior. If the agent finds a pattern that has a confidence level higher than a preset "tell-me" threshold, it can alert the user to the pattern and suggest the next action, then wait for confirmation to proceed. If the agent is confident above a second, higher "do-it" threshold, the agent can go ahead and autonomously perform the action. The user interface to the agent is a simple line drawing of a face in the corner of the screen. The face has different expressions, depending on the status of the agent—working, ready to offer a suggestion, unsure about what a user will do next, gratified that the user took a suggestion, surprised that the user rejected a suggestion, pleased to have anticipated an action, or confused to have failed to anticipate one. These expressions offer the user a quick and relatively unobtrusive way to see the status of the agent.

Because the agent learns, it may not be a good predictor initially. Users can get around this problem by explicitly instructing the agent with hypothetical situations and what to do, creating either default or hard and fast rules for the agent. In addition, the agent can collaborate with other mail agents. If the agent "does not have enough confidence in its prediction (confidence that is less than 'tell-me' threshold), it asks for help from other agents assisting other users with email" (Maes, 1994, p. 37). The mail agent is built to work with Eudora, a popular Macintosh mail program. The Media Lab reports that "results of. . .user tests are very encouraging. Users are eager to try out interface agents. . . .Users reported they felt comfortable delegating tasks to the agents" (p. 38). One interesting request from users was for a way to tell the agent to ignore mistakes. Overall, MIT's learning agents demonstrate the potential usefulness of knowledgable agents acting as information filters for user.

COACH is a teaching agent that Ted Selker described in a recent ACM issue (Selker, 1994, p. 92). COACH provides help to its user in learning how to program in Lisp. The COACH program keeps track of the user's actions and current context in order to offer help on the appropriate topic at the appropriate level of difficulty. The interface is entirely text-based—the agent's comments appear in a window above the user's work space. The user can also use a menu-based help function to actively seek advice on a particular topic. In a comparison of users of COACH and users of the regular menu help, Selker found that "the automatic adaptive help group utilized all available materials, felt more comfortable with Lisp, had a higher morale and wrote five times as many functions on average than the group with manual COACH help." (p. 98). He noted that COACH can be adapted to other knowledge domains with similar structures (i.e., syntax and semantics to master). COACH is a good example of how a fairly nonanthropomorphic agent can take advantage of context and observation of user patterns to perform a traditional human role with noteworthy success. In a domain where users expect little proferred help from the machine, even commentary with no dialogue can make a positive difference in learning.

Vellum's Drafting assistant is part of a CAD application with an intelligent drawing aid that helps the user construct figures with precise geometric relationships. As the user draws, the drafting assistant both notices and tries to anticipate when the user is approaching a geometric relationship of interest (for example, her line reaches or nears the midpoint of a circle). The agent communicates these anticipated relationships to the user with dotted lines and short text descriptions of the target relationship. If the user sees the agent find the relationship she wants, she simply releases the mouse and the agent draws the one depicted. The elegance of this design is that there is a tight dialogue going on between user and agent that enables the user to take quick advantage of the agent's domain knowledge without relinquishing control entirely (Tognazzini, 1992). This agent shows that agent techniques can be applied, without using natural language dialogue or anthropomorphism, to an ongoing operation.

*Autonomous agent (concrete and formal operations stages).* This type of agent can perform actions that conform to its own goals and respond to queries about those goals. It can adapt to individual users' responses and to patterns it detects through ongoing observation of users and can offer reliable advice about its areas of expertise, like the knowledgeable agent. The primary characteristic of this stage in agent development is the agent's independent agenda. This type of agent feels most like an independent individual when one interacts with it.

*Examples.* Julia the robot is a "Maas-Neotek robot" written by Michael Mauldin (for a more comprehensive description of Julia, see Foner, 1993). Julia travels within ongoing text-based interactive worlds on the internet. (These worlds, sometimes called *MUDs* (multiuser dimensions/dungeons), are virtual spaces that players can log into and "travel," using simple text commands. Players can also help shape the architecture of the spaces by building rooms and adding details. Players spend most of the time chatting with one another while in these worlds. For a more thorough description of these worlds, see Rheingold, 1993.) Julia is able to carry on reasonably lifelike text conversations with other users in MUDs; she keeps a map of the MUD she is in, and players can ask her how to get from one room to another. Julia also stores bits of conversation tagged by player, and can "play back" these bits on request. Players can ask Julia for statistics on the noisiest room, possible next destinations, and other information. This robot incorporates social skills that make it a better conversationalist. For example, if you whisper to it, it will whisper back. (Whispering is text that only goes between two users, and isn't seen by the other players that are present.) Julia was developed in part to perform in Turing competitions, so she has been given some highly anthropomorphic qualities. She can describe herself in several ways, knows where she is and what her gender is, and can carry on sophisticated dialogue, remembering references and events. She wanders MUDs gathering her data according to her own goals, interacting with people as they are encountered and approach her.

Julia is a good example of what we would label autonomous agents. She has her own goals and travels along, fulfilling her goals, interacting with users insofar as this coincides with her overall objectives. She is useful/enjoyable for the very reason that she has this autonomous quality, and is not tightly controlled or dependent on ongoing feedback from any one user.

## FIRST STEPS TOWARD GUIDELINES FOR AGENTS

This section attempts to give designers/developers some guidelines for introducing agents into the interface of applications/environment. Currently there are no "look and feel" guidelines for agency, and little quantitative research on what works. The following is an attempt to take a first cut at developing some guidelines that designers and programmers could use. The guidelines should be taken as suggestions meant to generate interest and thought, and encourage discussion and more research in the area.

### Why Use Agents at All?

As you can see from the examples in the previous section, agents can be used to perform tasks for users in ways that are far more intuitive and simple than other interface concepts would be. This is the most compelling reason to use any interface innovation—that it makes tasks easier and more pleasant for users. Why do agents make the interface more intuitive when used wisely? For one thing, they can focus the user on the most appropriate level of detail. Agents can take care of mundane or repetitive aspects of tasks, and can automate processes so the user doesn't have to know exactly how every step in the program works (like removing the need to know where every file is, or where something is located on the network). This way the user can concentrate on his or her part in executing a task—he/she is not forced to make every choice—only the important choices. This is a continuation of the steady trend in software toward more and more selective exposure of the workings of the program to the user.

Agents can also be used to perform tasks that the user does not want or cannot do for themselves. Huge bibliographic searches and filtering of internet newsgroup postings for specific information over a long period of time are the sorts of tasks for which an agent is ideal. Tasks that humans seek to delegate, so that the power of the computer, especially the power to move vast quantities of information, is well-used.

Also, agents take advantage of our natural tendency to attribute human qualities to objects we interact with (Laurel, 1990). Conscious design of the agent metaphor is a logical extension of the user's innate interpretive strategies, and makes the interface more seamless and natural. Calling a file finder an agent gives the user and the programmer guidelines for what to expect of the finder, and direction in creating a useful program.

Along the same lines, agents can act in human helper roles for users, providing a more natural and satisfying way to get assistance. A more human-like interaction may encourage exploration and learning in ways that pull-down help or command help simply cannot (Foner, 1993).

Finally, agents may add an entertaining social aspect to computing to balance out worktime. A well-crafted agent that is occasionally witty could be refreshing. Thinking of computers as being exclusively task-oriented neglects the proven value and ability of computers to function as entertainment. Agents could bring some of the fun from the entertainment world to work- or home-task computing, which may in fact make people more productive and convince a broader audience that computers are both useful and fun to use.

Some reasons for using agents:

1.  agents can provide assistance to users (offering advice, making sugges tions, training);
2.  agents can make users more productive/effective;
3.  if there are a number of cpu cycles on users' computers that are not being used, users can put agents to work for them and take advantage of these cycles;
4.  users can off load repetitive and/or mundane tasks to their agent;
5.  agents allow users to focus on the critical tasks and give agents knowledge of things that users should not need to know, such as loca tion of files or knowledge about the network;
6.  agents provide a good conceptual model to help deal with delegation and communication.

At the risk of stating the obvious, it is important to keep in mind when designing agents that they will never be as good at being people as people. People are terrific at information filtering, at teaching, and at social interaction (Ehrlich & Cash, 1994)—it is a hopeless endeavor to embark on a sort of Turing-test design strategy when emulating any human activity with an agent. The point is to leverage the human ability, not to try to faithfully reproduce it in the agent. This is why intelligent selection of abilities and representation are essential to good agent design.

## Concerns About Anthropomorphizing

One of the greatest concerns associated with using agents is that they encourage anthropomorphism, and thus both improper understanding of the computer's abilities and disappointment and frustration. We would like to suggest a separation of the notion of agents and explicitly human-looking agents. One can evoke many human qualities without necessarily encouraging anthropomorphizing. Agents like the Vellum drafting assistant make use of dialogue and feedback and adaptation without evoking exaggerated expectations or attributions from the user.

In addition, there is a long tradition of using human images in the arts to evoke emotions and reactions. Encouraging anthropomorphizing with human-like appearance of some sorts of agents may bring great power and richness to the user's interaction with some kinds of software.

Finally, some researchers suggest that people already unconsciously anthropomorphize computers (and anything else) that provides certain cues such as feedback or voice (Nass, Steuer, & Tauber, 1994). If this is in fact the case, the question then becomes not should we encourage anthropomorphizing, but what cues.should we use to elicit the amount of social response that will be useful in interacting with this agent?

The decision to evoke strong human qualities should certainly not be taken lightly. Users will rightly expect certain levels of interaction based on the promise of the agent's appearance in its context. This should not be confused with the decision to use the agent concept at all, though, and should not, we believe, deter explorations of the power of evoking human social behavior and expectations.

## Applying Human Qualities to Agents

Following are some questions to ask when designing agency into your application or environment.

*What is the agent supposed to do?*    What is the stated purpose of the agent? Various roles that an agent may fulfill are: (a) educator/trainer, (b) personal assistant/administrative assistant, (c) mail expert or domain expert, such as a CPA, and many others. The agent's purpose or task will have enormous consequences for what sorts of abilities it needs, and how it should be presented to the user.

One of the many design challenges when designing agents for use in the interface is making clear and obvious what abilities the agent might have or doesn't have (the limitations). To say that the agent is your mail agent, it would be fair for the user to assume that the agent should also be able to do whatever tasks they know how to do in mail. What would be even more helpful, is if the agent knows how to do things that the user cannot do or does not know how to do.

*How will the user talk to the agent?*    There are many ways in which a user could communicate with an agent: It could be command-line driven, offer the user dialogue boxes to fill in, record user actions and create macros, be designed to parse natural language, or recognize gestures. Each of these methods has benefits and problems associated with it.

Some of the obvious pros and cons for agents that are nothing more than dialog boxes or canned macros is that they are easy for the user to understand and take advantage of when it is needed. Command-line communication suffers from the same problem from which all command-line interfaces suffer; that is, the user must know the commands and how to communicate with the agent in order to get any help. Interfaces that allow users to have their actions recorded by the agent so they can be played back at any time are very nice for tedious and complex tasks, but they require that the user know how to do the task to begin with (and depending on the actual task that the user is trying to do, it could require that the task be edited in order to work properly). We found that natural language is the best method for making request because the user has a more natural form of communication with the agent and can make requ est for things that they do not understand or know how to perform. Although a natural language interface to the agent would be nice, it is quite complex and difficult to implement, and therefore not a pervasive form of interaction across platforms.

*How will the agent talk to the user?* What sort of feedback will your agent give the user? Does it explain errors when they occur? How does the user know when the agent has completed a task? Does the agent ask for clarification if it can't do a task as requested? Can it respond to a query, or does it only offer predetermined advice? It is important to match feedback sophistication to the user's input ability (Laurel, 1990).

*Who's in control?* The level of control a user has over the agent depends on how frequently he or she can give directions to the agent, and how meaningful

the choices are for the final product. It also depends on whether the agent will only do things at the moment, or will perform tasks asychronously. Perceived control also depends on whether the agent learns only when the user explicitly informs it, or learns by observation of the user's actions. Some agents may watch and merely advise; others might observe and then run off and execute a perceived goal without checking in. In general, the more control you design into the agent, the more carefully you should examine how comfortable the user feels with the agent's abilities and why (see that following).

*How confident is the user in delegating to the agent?* The level of delegation with which the user is comfortable is a complex factor, and depends on that amount of trust the user builds up based on the abilities of the agent and the riskiness of the tasks it performs (Foner, 1993). The level of delegation may shift over time, as the user sees the agent successfully perform tasks. Some of the things that you can do to help in this confidence building are to: (a) allow the user to see the task that is being performed instead of having it played in the background, and (b) give sufficient feedback to the user that the task has been completed and perhaps provide a log of agent activities for the day, week, or whatever timeframe makes sense. The more risky the task in terms of loss to the user (such as data loss), the more trust there must be for the user to delegate such a task to an agent. And with good reason. You may want to offer different thresholds of delegation to the user to adjust as desired (Maes, 1994).

*How human does the agent look?* How human does your agent appear (and sound)? Perhaps the answer for your application is "it doesn't sound or appear to be human at all." Applications such as Vellum present information and options to users by making suggestions, but the users don't associate those suggestions with a persona. All of the qualities in the "Looks" and "Personality" sections earlier in the chapter apply here. They are all aspects of humanness that can be used in agents, with good or ill effects, depending on context. In an entertaining or creative environment, you might want to use explicitly human-looking agents to give a social and emotional feel to the interface. In an information navigation framework, such qualities might be distracting and confusing. In any case, the more human-looking your agents are, the more users' expectations can rise about the ability of the agents to act like humans in other ways—such as engaging in conversation, obeying rules of politeness (Nass et al., 1994) and turntaking, etc. Thus human appearance shouldn't be added without forethought.

More specifically, the particular aspects you choose lead to expectations—a hyperrealistic image may lead to greater expectations than a highly cartoonish one. And you needn't assume that when you attach a persona to an agent it must be human. Agents might look like other animals, or like robots or imaginary creatures such as space aliens.

## Issues to Consider

Will users relate better to situated agents (mail agent and calendar agent) or a centralized agent? It is all in the presentation, of course, because a centralized agent may in fact communicate with lots of agents that have specific knowledge. The key question is how the user perceives the agent.

How can one inform users of agent capabilities—what the agent can do for the user? Match presentation to function—don't encourage expectations beyond what you can deliver with the depiction or selling of your agent. This doesn't mean that users can't build a rich relationship with a personification; they just need to get a clear understanding of the limits of what it can and cannot do, regardless of the code underneath the apparent intelligence. If you want to encourage a social bond, then anthropomorphize. If your focus is on a businesslike interaction and delegation, you might want to question the merit of elaborate anthropomorphizing (although isolated cues like a changing facial expression in a mail filter can give the user quick cues about the agent's status efficiently (like the MIT mail filter; Maes, 1994). Keep in mind that you can convey functionality with focused anthropomorphizing in the same way that icons convey functionality. However, gratuitous characterization could work against you if you aren't designing games or other environments where the point of the interaction is social or entertaining. Also keep in mind that an agent will be used over and over again, so idiosyncratic behaviors that might be funny once could become torture after the hundredth time.

One must also ask: What identity will an end user want for the agent? Will he or she prefer something that resembles a human or animal, or something that appears to be another, more sophisticated, piece of software? Think through the interaction: Because users will need to trust the agent in order to delegate, it is crucial to provide an adequate dialogue. If the agent flops to a halt due to bad syntax and doesn't offer any help about what to do next, the user will soon stop using the agent, and will grow to hate it. Use social interaction as a model— how do we repair communication when there is a misunderstanding? How do we interrupt politely? How do we make suggestions, and what are the cues we should heed to back off? (Clark & Brennan, 1991). Most forms of agent are all about the user relinquishing control of the computer for a time. This is NOT an intuitive or comfortable thing to do, and will not be taken lightly. It is important to users to understand exactly what the computer will and won't do, and to feel confident that it will check in with them if it is confused. When designing agents that offer advice, be sensitive to users who may not want advice all the time, or who may want to moderate the agent's level of advice. Give the user the ability to take back the reins as much as possible.

## CONCLUSION

Agency is a growing area of interest for interface design. Multitasking and complex networked interaction have taxed the direct manipulation approach—interface agents may offer a workable alternative or supplement. However, designing useful agents involves the consideration of a host of issues that aren't nearly as prominent when designing for direct manipulation. This chapter has attempted to offer the reader broad guidelines for considering these issues, from the perspective that agents are used to emulate human action, assistance, and communication. Designers will need to spend time thinking through which aspects of humanness they want to evoke for what purposes, and develop good mappings for users. We hope this chapter will spark further debate and experimentation with agent qualities.

## REFERENCES

Clark, H. H., & Brennan., S. E. (1991). Grounding in communication. In L.B. Resnick, J. M. Levine, & S. D. Teasley (Eds.), *Perspectives on socially shared cognition*. Washington, DC: APA Books.

Crowston, K., & Malone, T. W. (1988). Intelligent software agents. *Byte, 13*(13), 267-74.

Dent, L., Boticaria, J., McDermott, J., Mitchell, T., & Zabowski, D. (1992). A personal learning apprentice. In *Proceedings of the National Conference on Artificial Intelligence*.

Edmonds, E. A., Candy, L., Jones, R., & Soufi, B. (1994). Support for collaborative design: Agents and emergence. *Communications of the ACM, 37*(7), 41–47.

Ehrlich, K., & Cash, D. (1994). Turning information into knowledge: Information finding as a collaborative activity. In *Conference Proceedings of Processing of Digital Libraries '94*.

Foner, L. (1993). *What's an agent, anyway? A sociological case study*.

Foner, L., & Maes, P. (1994). Paying attention to what's important: Using focus of attention to improve unsupervised learning.

Gentner, D., & Nielsen, J. (1994). *The anti-Mac interface*. Abstract in CHI '95 Conference Companion.

Humans and others: The concept of agency and its attributions. *American Behavioral Scientist, 37*(6).

Kay, A. (1984). Computer software. *Scientific American, 251*(3).

Kay, A. 1990. User interface: A personal view. In B. Laurel (Ed.), *The art of human–computer interface design*. Reading, MA: Addison-Wesley.

Lashkari, Y., Metral, M., & Maes, P. (1994). Collaborative nterface agents. In *Proceedings of the National Conference on Artifi cial Intelligence*.

Laurel, B. (Ed.). (1990). *The art of human–computer interface design*. Reading, MA.: Addison-Wesley.

Laurel, B. (1993). *Computers as theater*. Reading, MA: Addison-Wesley.

Lebowitz, M. (1984). Creating charact ers in a story-telling universe. *Poetics, 13*, 171–194.

Lieberman, H. (1993). Mondrian: A teachable graphical editor. In A. Cypher (Ed.), Watch what I do: Programming by demonstration. Cambridge, MA: MIT Press.

Maes, P. (1994). Agents that reduce work and information overload. *Communications of the ACM, 37*(7), 31–40/

Maes, P., Darrell, T., Blumberg, B., & Pentland, S. (1994). Interacting with animated autonomous agents. In Working Notes AAAI Spring Symposium on `Believable Agents', Stanford, CA.

Maes, P., & Kozierok, R. (1993). Learning interface agents. In *Proceedings of the AAAI `93 Conference* . Cambridge, MA: MIT Press.

Nass, C., Steuer, J., & Tauber, E. R. (1994). Computers are social actors. In *CHI `94 Conference Proceedings*.

Nielsen, J. (1993). Noncommand user interfaces. *Communications of the ACM, 36*(4), 83–99.

Rheingold, H. (1993). The virtual community: Homesteading on the electronic frontier. Reading, MA: Addison-Wesley.

Sheth, B., & Maes, P. (1993). *Evolving agents for personalized information filtering.*

Shoam, Y. (1990). Agent-oriented programming. *Artificial Intelligence, 60*(1), 51–92.

Stearns, G. (1989). Agents and the HP NewWave application program interface. *Hewlett-Packard Journal, 40*, 32–37.

Sullivan, J. W., &  Tyler, S. W.  (Eds.). (1991). *Intelligent user interfaces.* New York: ACM Press.

Wood,  A.  Agent-Based  Interaction.  Paper  available  at  WWW  site :http://www.cs.bham.ac.uk/ ~amw/agents/index.html

# Chapter 4
# Toward Accessible
# Human-Computer Interaction*

Eric Bergman
*SunSoft*

Earl Johnson
*Sun Microsystems Laboratories*

## INTRODUCTION

In spite of the growing focus on user-centered interface design promoted by human-computer interaction researchers and practitioners, there remains a large and diverse user population that is generally overlooked: users with disabilities. There are compelling legal, economic, social, and moral arguments for providing users with disabilities access to information technologies. Although we will touch on some of those arguments here, the primary purpose of this chapter is to provide a broad overview of the software human-computer interaction (HCI) issues surrounding access to computer systems.

The needs of users with disabilities are typically not considered during software design and evaluation. Although there are many plausible explanations for this omission, we are inclined to believe that much of the problem simply stems from lack of awareness. Until recently, there was little contact between HCI organizations and the sizable community of people with disabilities (McMillan, 1992). As a result, many software designers are not aware of the needs of users with disabilities. We hope that this chapter will serve to increase awareness of the scope and nature of these needs, and to stimulate interest in research and implementation of systems that enable access to information technologies by users with disabilities.

Designing software that takes the needs of users with disabilities into account makes software more usable for all users: people with disabilities who use assistive technologies and those who use systems "off the shelf," as well as users without any significant disabilities. Considerable literature already exists that discusses how people with disabilities can use assistive hardware and software to interact with computers. For this reason, we provide relatively brief

*The authors would like to thank Ellen Isaacs, Beth Mynatt, Mark Novak, and Will Walker for their valuable suggestions and comments on this chapter.

coverage of assistive technologies here.[1]

In this chapter, we define and discuss accessibility, discuss accessibility design issues, provide a broad outline of the capabilities and needs of users with various disabilities, present guidelines for increasing application accessibility, and discuss future directions for improving accessibility.

## THE RELEVANCE OF ACCESSIBILITY

### What is Accessibility?

Providing accessibility means removing barriers that prevent people with disabilities from participating in substantial life activities, including the use of services, products, and information. We see and use a multitude of access-related technologies in everyday life, many of which we may not recognize as disability related when we encounter them. The bell that chimes when an elevator is about to arrive, for example, was designed with blind people in mind (Edwards, Edwards, & Mynatt, 1993). The curb cut ramps common on street corners in the United States were introduced for wheelchair users (Vanderheiden, 1983). Accessibility is by definition a category of usability: software that is not accessible to a particular user is not usable by that person. As with any usability measure, accessibility is necessarily defined relative to user task requirements and needs. For example, a telephone booth is accessible (e.g., usable) to a blind person, but may not be accessible to a person using a wheelchair. Graphical user interfaces are not very accessible to blind users, but relatively accessible to deaf users.

Vanderheiden (1991) made a distinction between "direct" access and access through add-on assistive technologies. He described direct access as "adaptations to product designs that can significantly increase their accessibility" (p. 2). A major advantage of this approach is that large numbers of users with mild to moderate disabilities can use systems without any modification. Examples of direct access include software keyboard enhancements included with X Windows, OS/2, and the Macintosh (see Table 2).

Assistive access means that system infrastructure allows add-on assistive software to transparently provide specialized input and output capabilities.[2] For

---

[1]For more detailed information on assistive technologies see Lazzaro (1993), Church and Glennen (1992), Brown (1992), and Casali and Williges (1990).

[2]By *infrastructure* we mean an environment's standard set of application program interfaces (APIs). These are low- and high-level software routines used to build applications (e.g., Macintosh Toolbox, MS Windows API, Motif, and Xlib to name a few).

example, screen readers (see Table 3) allow blind users to navigate through applications, determine the state of controls, and read text via text to speech conversion. On-screen keyboards replace physical keyboards, and head-mounted pointers replace mice. These are only a few of the assistive technologies users may add on to their systems.

We claim that in order to truly serve users with disabilities, accessibility must mean more than simply providing "direct" access through assistive technologies bundled with system software, and more than providing the capability to add such assistive technologies. It also must mean designing application user interfaces that are easier to use for users with disabilities as well as users "with out" disabilities by taking their needs into account when system and application software is designed.

## Broad Benefits of Accessibility

Accessibility provides benefits for a wide range of people—not only for those with disabilities. Before curb cut ramps were placed on sidewalks, for example, it was difficult or impossible for people in wheelchairs to cross a street. In addition, curb cuts have benefited people pushing baby carriages and shopping carts as well as those on bicycles and roller blades. Vanderheiden (1983) suggested that our society is laying down electronic equivalents to sidewalks. These "electronic pathways," he argued, must include electronic "curb cuts."

Like physical curb cuts, electronic curb cuts provide benefits for the larger population as well as users with disabilities. Systems that allow use of keyboard shortcuts instead of the mouse increase efficiency of sighted users as well as providing access for blind users or users who have disabilities that affect mouse control. Users of portable computers or people in open offices may not be able to use or hear sounds, but they can use visual cues, as can hearing impaired users. Users who must keep their eyes on their task (e.g., aircraft mechanics) can benefit from systems that interact through voice rather than vision, as can users with visual disabilities.

In the near future, when the power of computing is available on noisy factory floors, in cars hurtling down expressways, and from devices stuffed in our pockets, the relevance of accessibility looms larger. As users and designers, we will soon deal with environment and task demands in which systems must deliver computing power to people who will be unable to use their hands, eyes, or ears to interact with computers. People working on accessibility have been tackling such design issues for years, typically with little or no input from HCI specialists. Clearly there needs to be a better communication between the disability access and HCI communities.

There are ample incentives for fostering a connection between design for access and design for "general audiences." Research on communication devices for the deaf led to development of the telephone, whereas development of an easy to use "talking book" for the blind led to the cassette tape (Edwards, Edwards, et al., 1993). HCI theory and practice can benefit from better understanding of the difficult issues that are already being addressed in the disability domain. Newell and Cairns (1993) cited designers who thought they had created novel interfaces, which were actually reinventions of disability access technologies such as foot-operated mice and predictive dictionaries.[3] As McMillan (1992) suggested, cooperation among these different communities will require better communication among professionals in the fields of rehabilitation education and HCI.

## Economic and Social Impact

A common argument is that computer accessibility is too costly, but in reality inaccessible systems may cost much more. Government statistics show that there is a growing market for accessible computer products. Approximately 43 million Americans have a disability of some type (Perritt, 1991). According to Elkind (1990), about 15% of the population has a disability "severe enough to interfere with work or otherwise constitute a serious handicap," whereas Vanderheiden (1990) pointed out that over 30% of the people with disabilities who want to work are not employed.

There are human costs that are harder to measure than numbers from tax tables or percentages of people employed. As a user with low vision told us, "I don't believe in telling peers or management [that] I can't read it on the computer so I can't do my work." We have spoken to blind users who, because they are unable to read text that they have typed into word processing applications, have secretaries or coworkers type for them. We have spoken to users with mobility impairments who could not use job-critical applications because they were not accessible from the keyboard. Our experience suggests that a significant proportion of users with disabilities who are employed are not working at full productivity due to poor software accessibility.

## Legal Requirements

In recent years, new laws have created incentives for the development of access to computer technology (see Casali & Williges, 1990; Lazzaro, 1993; McCormick, 1994). The most important of these laws, for the purposes of our discussion, are Section 508 of the 1986 Federal Rehabilitation Act and the 1992 Americans with Disabilities Act (ADA).

---

[3]Predictive dictionaries speed typing by predicting words as the user types them and offering those words in a list for the user to choose. Originally intended for users with movement-related disabilities, predictive dictionaries are becoming popular for users with RSI and as a way to boost typing speed.

**Key aspects of Section 508:**
- Applies to office computer purchases made by the Federal government
- Winning contract proposals must include solutions for employees with disabilities

**Key aspects of the ADA:**
- Applies to corporate entities with 15 or more employees
- Requires "reasonable accommodation" for employees with disabilities

One of the driving forces behind these laws has been the government's economic and social interest in retaining and recruiting people with disabilities. In fact, the government employs over 151,000 people with disabilities (McCarthy, 1994). The ADA was seen as a way to make the private sector a partner in hiring and retaining employees with disabilities, reducing demand for government assistance. The growing market for accessible technologies means that accessibility can be an important selling point for system and application software in both commercial and federal markets.

## DESIGNING FOR DISABILITIES

Arguments typically leveled against designing for users with disabilities include the claims that costs are too high, and the benefits serve too small a market (Glinert & York, 1992). These arguments should sound familiar to HCI practitioners, who have historically faced initial resistance or even opposition to the introduction of HCI processes into product development. Just as organizational understanding of the design process must be changed for HCI to be accepted, so too must the standard HCI conceptualization of "the user" change to recognize the needs of people with disabilities.

### Ranges of User Capabilities

The traditional view of people "having a disability" or "not having a disability" is overly simplistic. All users have a range of capabilities that varies across many dimensions depending on the user and his or her life stage, task, and environment. As more of the population approaches their middle 40s, there are increasing numbers of users peering through bifocals at their screens. A nontrivial portion of the population experiences some degree of hearing loss, and may not always notice software alert sounds. As we age, more of us will develop age-related disabilities—25% by age 55, jumping to 50% at age 65 (Vanderheiden, 1990).

In addition to normal consequences of aging, people may experience sudden temporary or permanent changes in capabilities at any time in their lives. If a computer user falls and breaks her wrist, she will spend several weeks or more with much the same keyboard capabilities as many people with spinal cord injuries or missing limbs. Every user's ability to interact with systems varies over short as well as long periods of time. A sprained wrist can restrict use of a mouse. A late night working or a lost contact lens can transform small fonts into a suddenly challenging experience. Any user who does not currently have a disability could someday have a stroke, car accident, or other event resulting in a temporary or permanent disability.

In fact, a significant number of user requirements for people with disabilities apply to almost any user, given the right circumstance or task context (Edwards, Edwards, et al, 1993; Newell & Cairns, 1993). Whether a user's hand is broken, painful due to repetitive strain injury, or permanently paralyzed, there are similar needs. Whether someone is unable to look at a screen because he is driving, or cannot see a screen because he is blind, the user requirements have much in common. Whether a user does not hear because she is talking to somebody on the phone, paying attention to her task, working in a noisy environment, or happens to be deaf is less important than the fact that users in these contexts need alternate sources of information. As McMillan (1992) observed, "From the point of view of a computer, all human users are handicapped" (p. 144).

**Who is "the User?"**

Who is "the user"—that familiar catchphrase we encounter in papers, conference sessions, and design sessions? Does it include users with disabilities? Does "the user" include a programmer we visited? She was diagnosed with muscular dystrophy in her early 20s. This condition, which results in progressive loss of muscular strength, means that she works from her motorized wheelchair, and is unable to sit upright for more than a brief time. As a result, she works in a reclined position, leaning back almost horizontally. Her vision problems limit the amount of time she can focus on the screen, and her muscular weakness prevents her from handling paper manuals.

Does "the user" include a secretary we interviewed? She has no vision in one eye and "tunnel vision" in the other and prepares documents using a standard PC and screen magnification software. Sometimes she is unable to tell the difference between old and new email messages, because her mail application uses color to distinguish old from new. Like many users with low vision, she has problems working with columns, because it is difficult for her to see if text is aligned.

Does "the user" include a writer we know who took several months off from work when she developed tendonitis? She was not able to type or use a mouse for more than a few minutes, nor was she able to lift heavy objects, including manuals. Although her condition improved significantly over time, for several months she had a disability which affected her work significantly.

Does "the user" include a computer support technician we met? He has cerebral palsy, and is able to use only his left hand. There are keyboard-assistive applications that would help speed his typing, but his work requires moving from computer to computer installing and troubleshooting Microsoft Windows software, which, at this writing, has no *built-in* access features.[4] It is not practical for this user to install an assistive keyboard application for each consultation, so he instead stretches his single usable hand wide to reach and press multiple keys when he needs control keys and capital letters. Fortunately for him, and his employer, he has a large hand.

These users and many others have told us that their needs are not being met by current computer systems. Users with physical disabilities complain about applications that cannot be controlled from the keyboard. Users with low vision describe software that does not allow them to adjust the color to make text legible. Blind users complain about documentation that is not accessible because it is not available on paper in braille or on the computer as plain text (which is required for screen-reading applications).

Access problems are not confined to users who have a "classic" disability. As they age, users who would claim they have no disability find that screens become more difficult to read and sounds become more difficult to hear. Users who break an arm, sprain a wrist, lose a contact lens, require bifocals, or develop repetitive stress injuries suddenly find that computer systems do not take their needs into account.

## Limited View of the User

A common argument goes something to the effect that software engineers typically design for themselves, whereas HCI professionals follow a process based on understanding user characteristics, needs, tasks, and environments. In spite of this claim to the domain of user needs, we argue that most HCI research, literature, and practice holds forth a relatively limited view of who constitutes "the user." Most of the people we have discussed as examples of users with disabilities would not fit in this limited view.

Nielsen's (1993) Usability Engineering, for example, discussed the user in a section titled "Categories of Users and Individual Differences." Nielsen focused on user experience with computers in general, the system in particular,

---

[4]Note that this is scheduled to change in "Windows 95." As demonstrated in this case, add-on software alone does not meet the needs of a mobile user.

and the task at hand. He noted that "users also differ in other ways than experience" and went on to list such attributes as age, gender, spatial memory, and learning style. Users with disabilities are not mentioned, in spite of the explicit focus on "Categories of Users." In fact, disabilities are only mentioned in a few brief sentences in the entire book.

We use Nielsen's book only as an example of the extent to which disability issues are ignored by mainstream literature. Although there is a growing body of work on accessibility within the HCI community, it has not been generally recognized in standard texts or in work that is not explicitly focused on disability issues. Users with disabilities are simply not "on the radar screen" of mainstream HCI.

## Universal Design

In recent years, as the notion of accessibility has been broadened to encom pass much more than design for people with disabilities, the concept of "universal design" has gained visibility. Traditional design has focused on filling the needs of the "average" person , with the assumption that design for the average provides for the needs of most. The universal design argument is that designing for the "average" is by definition exclusionary, because the "average" user is a fictitious construct.

Attempts to design for this fictitious "average" user may not account for effects of multiple individual differences. Tognazzini (1992, p. 74) related an anecdote from Blake (1985) illustrating the pitfalls of ignoring such overlapping differences:

> Several years ago, the Air Force carried out a little test to find out how many cadets could fit into what were statistically the average-size clothes. They assembled 680 cadets in a courtyard and slowly called off the average sizes—plus or minus one standard deviation—of various items, such as shoes, pants, and shirts. Any cadet that was not in the average range for a given item was asked to leave the courtyard. By the time they finished with the fifth item, there were only 2 cadets left; by the sixth, all but one had been eliminated.

The Universal Design philosophy emerges from a recognition of the idea central to this story—that there is no average user. Universal design targets the broadest possible range of user capabilities. Examples of products that embody this theme include automatic doors, remote control thermostats, and Velcro. Using no assistive technology, people who were previously unable to open a door, operate a thermostat, or tie their shoes are able to perform these tasks, whereas "the rest of us" find these tasks easier as well. Proponents of universal design do not assume that all users will be able to use all designs, but instead argue that by redefining our definition of the user, a much wider range of users can be accommodated without significant extra effort (Vanderheiden, 1992a).

## Watch out for that Ramp

We claim that Universal Design is a worthy goal, but it is important to acknowledge that there are complex customization-related HCI issues that must be resolved before it can be achieved with computers. In discussing user interface design, Lewis and Rieman (1994) wrote, "Experience shows that many ideas that are supposed to be good for everybody aren't good for anybody." We agree that in human–computer interaction, as in much of life, what is "good for you" is not always "good for me."

An example of this principle in action was illustrated to us by a colleague who caught her foot at the base of a wheelchair ramp and tripped. The resulting fall produced injuries that included a sprained wrist and numbness in one hand. The injuries could easily have been more severe. The irony of a person acquiring a temporary or perhaps permanent disability because of an artifact designed to help people with disabilities strikes us as an appropriate alert that there may be stumbling blocks on the path to Universal Design.

One computer-related stumbling block is apparent in considering a simplified scenario of a public information kiosk. If we assume blind users must have access, then it becomes important to provide voice and sound output. There may be a tone when a control has been activated, and voice output to provide information about graphics and video displayed on screen. If a deaf user steps up to the kiosk, she will need visual notifications such as closed captions and visual alerts as alternatives to voice and sound alerts. If a user with no significant disability steps up to the kiosk, how will it interact with her? Surely she will not wish to deal with a cacophony of voice descriptions, closed captions, beeping dialogs and flashing screens?[5]

Environments are needed that allow users to tailor system input and output modalities to their capabilities and preferences. Recent research has suggested that information can be represented in a form abstracted from the particulars of its presentation (Blattner, Glinert, Jorge, & Ormsby, 1992; Fels, Shein, Chignell, & Milner, 1992). The technical solution of providing multiple redundant interface input and output mechanisms is not, in itself, sufficient to resolve conflicting user needs. In the absence of any means for intelligently designing and customizing their use, highly multimodal interfaces could lead to as many usability problems as they resolve, causing some users to trip over features

---

[5]This scenario is, of course, an oversimplification. Future systems are likely to adapt themselves to work intelligently with each user—perhaps based on a stored profile. How such systems become configured and interact with users remains an interesting HCI question.

designed to help other users. Determining how users will interact with such systems is a challenging HCI issue.

### Include People with Disabilities in the Design Process

It is in the design and evaluation of operating systems and desktop environments that designing for people with disabilities is most critical. Without the appropriate system software infrastructure, no amount of effort on the part of application developers can improve the accessibility of applications. As we discuss later, large gains in the accessibility of computer systems ultimately depend on improvements in software infrastructure.

On the other hand, there are many ways to improve the accessibility of applications within the constraints of current systems. Perhaps the most obvious way to enhance accessibility is to consider the needs of people with disabilities in all stages of the design process, including requirements gathering, task analyses, usability tests, and design guidelines. Other strategies include evaluating the usability of software in conjunction with popular assistive technologies, and testing under simulated disability conditions (e.g., unplug the mouse, turn off the sound, and use a screen reader with the monitor turned off). Note that none of these approaches are substitutes for testing with users. Simulation does not realistically represent the rich contexts and needs of users with disabilities. On the other hand, it is better than not testing accessibility at all.

Usability testing with even one user from each of the general disability categories we discuss in this chapter can have significant benefits for all users, not only those with disabilities. Depending on their disability, users can be especially affected by usability defects. Low-vision users are sensitive to font and color conflicts, as well as problems with layout and context. Blind users are affected by poor interface flow, tab order, layout, and terminology. Users with physical disabilities affecting movement can be sensitive to tasks that require an excessive number of steps or wide range of movement. Usability testing with these users can uncover usability defects that are important in the larger population.

## THE ROLE OF GUIDELINES AND APPLICATION PROGRAM INTERFACES

### Design Guidelines

Although large leaps in accessibility await improvements in the design of operating systems and desktop environments, there is much that can be done to improve access within the constraints of current systems. The extent to which applications follow design guidelines, for example, can have a disproportionate affect on people with disabilities. User confusion about how to perform tasks is always a problem, but such problems become magnified for users who use alternative input and output devices or who require extra steps o r time to navigate, activate controls, or read and enter text.

Keyboard mapping guidelines, for example, are especially important for those users who have movement impairments or are blind. Many of these users employ assistive software that assumes applications will use standard keyboard mnemonics and accelerators. Keyboard inconsistencies that are annoying to users without disabilities can become major roadblocks to users with disabilities.

Blind users of graphical user interfaces are especially affected by arbitrary violations of design guidelines with respect to application layout, behavior, and key mappings. These users interact with their systems through keyboard navigation, and use one-line braille displays and/or voice synthesis with screen reader software to read screen contents. Unlike a sighted user who can selectively scan and attend to screen elements in any order, blind users with screen readers move through a relatively linear presentation of screen layout. If that layout deviates significantly from guidelines, it can be especially difficult for a blind user to understand what is happening.

Because there are always exceptions, new situations, and missing guidelines, interface designers often have to violate guidelines or invent their own guidelines. In these cases, it is important that designers consider how any given violation might affect usability for users with disabilities. Unfortunately, most interface style guides were not written taking users with disabilities into account. For this reason, we have provided a set of general design guidelines in this chapter that can be applied across a variety of environments.

## The Application Program Interface (API)

Just as guidelines specify standard methods for interacting with an interface, various user interface application program interfaces (APIs) specify standard methods for applications to interact with each other and the system. API's are the low and high-level software routines used to build applications. Every software environment provides standard API functions to support activities such as reading characters from the keyboard, tracking position of the pointer, and displaying information on the screen.[6]

Assistive software and hardware mediates the communication between users and applications, making it particularly sensitive to cases where standard API functions are not used. This sensitivity occurs because assistive software

---

[6]Common APIs include the Macintosh Toolbox, the MS-Windows API, and Motif.

such as screen readers and speech recognition monitor the state and behavior of applications partly by tracking their use of API functions. For this reason, applications that do not use standard API calls have the potential to create serious usability problems for people with disabilities. If an application performs a common function (e.g., reading a character from the screen) without using standard API calls, assistive software may not know that an event has occurred, and consequently the user may not be able to use the application. For the same reason, a nonstandard interface component (e.g., a custom control) can cause access problems because chances are good that assistive technologies will not be able to recognize it or interact with it.

The API problem is a complicated issue, and beyond the scope of this chapter to address in detail, but it is important that project teams discuss and weigh the tradeoffs inherent in departures from APIs. Reasons for not using standard API functions include:

o It is not possible to implement the desired functionality using a high level API alone, because it does not support required user interface features;

o The high-level API calls do not yield desired performance;

o Developers are more familiar with lower level APIs that existed prior to the development of higher level APIs. In this case, developers may be more efficient or comfortable implementing existing features using the features they already know.

There are no simple solutions to the API problem. Where possible, engineers need to use APIs above the level in which assistive software connects to the system. If using a standard API constrains application functionality or performance, the reality of product or task requirements often means that it cannot be used. In cases where developers are unfamiliar with high-level APIs, training may help. When an interface feature is not supported by an API, it is rarely practical to drop that feature. In cases where an interface has features that clearly circumvent basic API functions, every attempt should be made to insure that the tasks these features support can also be accomplished using other features. In the future, access problems related to API support can be reduced by environments whose infrastructure provides built-in support for access.

## ABOUT DISABILITIES: BACKGROUND AND DESIGN GUIDELINES

In this section, we discuss some of the needs, capabilities, and assistive technologies used by people with disabilities, and provide guidelines for improving application accessibility. The brief descriptions in this section do not constitute complete coverage of the wide range of disabilities, capabilities, needs, and individual differences across the population of people with disabilities—we have focused on providing a broad introduction to visual, hearing, and physical disabilities. Users with cognitive, language, and other disabilities may have needs in addition to those discussed in this chapter.[7]

Use of assistive technologies varies across users and tasks. Our discussion of assistive technologies is not comprehensive, but it does cover many commonly used software and hardware solutions. In reading this section it is important to remember that as with all users, the needs of those with disabilities vary significantly from person to person. Many users with disabilities do not use assistive technologies, but can benefit from small design changes. Other users have significant investments in assistive technologies, but they too can benefit from software that better responds to their interaction needs.

## About Physical Disabilities

Physical disabilities can be the result of congenital conditions, accidents, or excessive muscular strain. By the term "physical disability" we are referring to disabilities that affect the ability to move, manipulate objects, and interact with the physical world. Examples include spinal cord injuries, degenerative nerve diseases, stroke, and missing limbs. Repetitive stress injuries can result in physical disabilities, but because these injuries have a common root cause, we address that topic later under its own heading.

Many users with physical disabilities use computer systems without add-on assistive technologies. These users can especially benefit from small changes in interface accessibility. As a case in point, we recently met a manager with Cerebral Palsy who uses a standard PC and off-the-shelf Windows business productivity applications, but navigates almost exclusively via the keyboard because his poor fine-motor coordination makes a mouse or trackball difficult to use. When a pointing device becomes necessary, he uses a trackball. As he explained it to us, "I hate the mouse."

Some users with physical disabilities use assistive technologies to aid their interactions (see Tables 1 and 2). Common hardware add-ons include alternative pointing devices such as head tracking systems and joysticks. The MouseKeys keyboard enhancement available for MS Windows, Macintosh, and X Windows-based workstations allows users to move the mouse pointer by pressing keys on the numeric keypad, using other keys to substitute for mouse button presses. Because system-level alternatives are available, it is not necessary for *applications* to provide mouse substitutes of their own. The problem of the mouse is a good example of the kind of generic issue that must be addressed at the system rather than application level.

Unfortunately, the MouseKeys feature is often time-consuming in comparison to keyboard accelerators, because it provides relatively crude directional

---

[7]We believe that in many cases consideration of the issues discussed here will address many of the needs of these users. See Vanderheiden (1992a) and Brown (1989) for more information on other disabilites.

Table 1. Assistive Technologies for Physical Disabilities and RSI

| Assistive Technology | Function Provided |
|---|---|
| Alternate Pointing Device | Gives users with limited or no arm and hand fine motor control the ability to control mouse movements and functions. Examples include foot operated mice, head-mounted pointing devices and eye-tracking systems. |
| Screen Keyboard | On-screen keyboard which provides the keys and functions of a physical keyboard. On-screen keyboards are typically used in conjunction with alternate pointing devices. |
| Predictive Dictionary | Predictive dictionaries speed typing by predicting words as the user types them, and offering those words in a list for the user to choose. |
| Speech Recognition | Allows the user with limited or no arm and hand fine motor control to input text and/or control the user interface via speech. |

Table 2. Keyboard Enhancements

| Feature | Function Provided |
|---|---|
| StickyKeys | Provides locking or latching of modifier keys (e.g., Shift, Control) so that they can be used without simultaneously pressing the keys. This allows single finger operation of multiple key combinations. |
| MouseKeys | An alternative to the mouse which provides keyboard control of cursor movement and mouse button functions. |
| RepeatKeys | Delays the onset of key repeat, allowing users with limited coordination time to release keys. |
| SlowKeys | Requires a key to be held down for a set period before keypress acceptance. This prevents users with limited coordination from accidentally pressing keys. |
| BounceKeys | Requires a delay between keystrokes before accepting the next keypress so users with tremors can prevent the system from accepting inadvertent keypresses. |
| ToggleKeys | Indicates locking key state with a tone when pressed, e.g., Caps Lock. |

control. For tasks requiring drag and drop or continuous movement (e.g., drawing), MouseKeys is also inefficient. On the other hand, because current systems are designed with the implicit assumption that the user has a mouse or equivalent pointing device, many tasks require selecting an object or pressing a control for which there is no keyboard alternative. In these cases, MouseKeys provides an option. It is clear that future operating environments need to offer effective alternatives for users who may not use a pointing device.

It is important that applications provide *keyboard access* to controls, features, and information in environments that have keyboard navigation.[8] Comprehensive keyboard access helps users who cannot use a mouse. Many environments allow users to use tab and arrow keys to navigate among controls in a window, space bar and enter to activate controls, and key combinations to move focus across windows. In some cases, extra engineering may be required to ensure that these features work in all areas of an interface.

In addition to keyboard navigation, keyboard accelerators and mnemonics are also helpful for users with physical disabilities (as well as blind and low-vision users). Whenever practical, commonly used actions and application dialogs should be accessible through buttons or keyboard accelerators. Unfortunately few of the standard accelerator sequences were designed with disabilities in mind. Many key combinations are difficult for users with limited dexterity (e.g., in Motif, holding down Alt–Shift–Tab to change to the previous window). Nonetheless, use of key mapping consistent with guidelines for the local application environment not only speeds use of applications for users with movement difficulties, but it also increases the effectiveness of alternate input technologies such as speech recognition. Assistive technologies often allow users to define macro sequences to accelerate common tasks. The more keyboard access an application provides, the greater the user's ability to customize assistive technology to work with that application.

## About Repetitive Strain Injuries (RSI)

Perhaps the fastest increasing disability in today's computerized workplace is repetitive strain injury (RSI). The Occupational and Health Safety Administration reported that 56% of all workplace injuries reported during 1992 were due to RSI, up from 18% in 1981 (Furger, 1993). RSI is a cumulative trauma disorder that is caused by frequent and regular intervals of repetitive actions. Repetitive strain injuries include tendonitis and carpal tunnel syndrome, although other types of injuries also occur. Symptoms of computer-based RSI include headaches, radiating pain, numbness, tingling, and a reduction of hand function. For computer users, mouse movements and typing may be causes or contributors to RSI.[9] Sauter, Schleifer, and Knutson (1991) found that repetitive use of the right hand among VDT data entry operators was a factor in causing RSI. They suggested that a change to "more dynamic tasks" could help reduce the likelihood of RSI. In general, users should be given the choice of

---

[8]The lack of keyboard control navigation in some user interface environments is an accessibility problem that needs to be addressed by the designers of these environments.

[9]We only discuss *software* strategies for reducing RSI in this section. Note that there is a wide variety of commercial ergonomic keyboards and alternative input devices aimed at reducing RSI.

performing a task using a variety of both mouse and keyboard options. For custom applications involving highly repetitive tasks, consider providing automatic notification for users to take breaks at regular intervals if there is no such capability at the system level.

Frequently repeated keyboard tasks should not require excessive reach or be nested deep in a menu hierarchy. We once met a customer-support representative who had an RSI-related thumb injury. One of her most common tasks—changing screens to register information from phone calls—encouraged an extremely wide stretch of her left thumb and forefinger in order to press a control and function key simultaneously. This thumb eventually required surgery.

The needs of users who have RSI overlap significantly with the needs of users with other types of physical disabilities. Assistive technologies such as alternate pointing devices, predictive dictionaries, and speech recognition can benefit these users by saving them keystrokes, reducing or eliminating use of the mouse, and allowing different methods of interacting with the system.

## About Low Vision

Users with low vision have a wide variety of visual capabilities. According to Vanderheiden (1992a), there are approximately 9–10 million people with low vision. For the purposes of this chapter, consider a person with low vision to be someone who can only read print that is very large, magnified, or held very close.

We recently met a user who can read from a standard 21" monitor by using magnifying glasses and the largest available system fonts. His colleague down the hall must magnify text to a height of several inches using hardware screen magnification equipment. A secretary we met has "tunnel vision," and can see only a very small portion of the world, as though she were "looking through a straw." In the region where she *can* see, her low acuity requires her to magnify text so that only a portion of her word-processing interface is visible on screen at any one time, reducing her view of the interface to only a single menu or control at a time. These are only a few of the wide variety of low-vision conditions.

The common theme for low-vision users is that it is challenging to read what is on the screen. All fonts, including those in text panes, menus, labels, and information messages, should be easily configurable by users. There is no way to anticipate how big is big enough. The larger fonts can be scaled, the more likely it is that users with low vision will be able to use software without additional magnification software or hardware. Although many users employ screen magnification hardware or software to enlarge their view, performance and image quality are improved if larger font sizes are available prior to magnification.

A related problem for users with low vision is their limited field of view. Because they use large fonts or magnify the screen through hardware or software, a smaller amount of information is visible at one time. Some users have tunnel vision that restricts their view to a small portion of the screen, whereas others require magnification at levels that push much of an interface off screen.

A limited field of view means that these users easily lose context. Events in an interface outside of their field of view may go unnoticed. These limitations in field of view imply that physical proximity of actions and consequences is especially important to users with low vision. In addition, providing redundant audio cues (or the option of audio) can notify users about new information or st ate changes. In the future, operating environments should allow users to quickly navigate to regions where new information is posted.

Interpreting information that depends on color (e.g, red = stop, green = go) can be difficult for people with visual impairments. A significant number of people with low vision are also unable to distinguish among some or any colors. As one legally blind user who had full vision as a child told us, his vision is like "watching black-and-white TV." In any case, a significant portion of any population will be "color blind". For these reasons, color should never be used as the only source of information—it should provide information that is redundant to text, textures, symbols, and other information.

Some combinations of background and text colors can result in text that is difficult to read for users with visual impairments. Again, the key is to provide both redundancy and choice. Users should also be given the ability to override default colors, so they can choose the colors that work best for them.

## About Blindness

There is no clear demarcation between low vision and true blindness, but for our purposes, a blind person can be considered to be anybody who does not use a visual display at all. These are users who read braille displays or listen to speech output to get informat ion from their systems (see Table 3).

Screen-reader software provides access to graphical user interfaces by providing navigation as well as a braille display or speech-synthesized reading of controls, text, and icons. The blind user typically uses tab and arrow controls to move through menus, buttons , icons, text areas, and other parts of the graphic interface. As the input focus moves, the screen reader provides braille, speech, or nonspeech audio feedback to indicate the user's position (see Mynatt, 1994). For example, when focus moves to a button labeled "search," the user might hear the words "button—Search," or when focus moves to a text input region, the user might hear a typewriter sound. Some screen readers provide this kind of information only in audio form, whereas others provide a braille display (a series of pins that raise and lower dynamically to form a row of braille characters).

Blind users rarely use a pointing device, and, as discussed earlier, typically depend on keyboard navigation. A problem of concern to blind users is the growing use of graphics and windowing systems (Edwards, Mynatt, &

Rodriguez, 1993). The transition to window-based systems is an emotional issue, evoking complaints from blind users who feel they are being forced to use an environment that is not well-suited to their style of interaction. As one blind user put it, "This graphics stuff gives sighted people advantages . . . it's user friendly . . . all this makes it user unfriendly for us [blind people]."

Although blind users have screen reading software that can read the text contents of buttons, menus, and other control areas, screen readers cannot read the contents of an icon or image. In the future, systems should be designed to provide descriptive information for all non-text objects. Until the appropriate infrastructure for providing this information becomes available, there are some measures that may help blind users access this information. Software engineers should give meaningful names for user interface objects in their code. Meaningful names can allow some screen reading software to provide useful information to users with visual impairments. Rather than naming an eraser graphic "widget5," for example, the code should call it "eraser" or some other descriptive name, that users will understand if spoken by a screen reader.

Without such descriptive information, blind or low-vision users may find it difficult or impossible to interpret unlabeled, graphically labeled, or custom interface objects. Providing descriptive information may provide the only means for access in these cases. As an added selling point to developers, meaningful widget names make for code that is easier to document and debug.

In addition to problems reading icons, blind users may have trouble reading areas of text that are not navigable via standard keyboard features. In OpenWindows and MS Windows, for example, it is not possible to move the

Table 3. Assistive Technologies for Low Vision and Blind Users

| Assistive Technology | Function Provided |
|---|---|
| Screen Reader Software | Allows user to navigate through windows, menus, and controls while receiving text and limited graphic information through speech output or braille display. |
| Braille Display | Provides line by line braille display of on-screen text using a series of pins to form braille symbols that are constantly updated as the user navigates through the interface. |
| Text to Speech | Translates electronic text into speech via a speech synthesizer. |
| Screen Magnification | Provides magnification of a portion or all of a screen, including graphics and windows as well as text. Allows user to track position of the input focus. |

focus to footer messages. If this capability were built into these environments, then blind users could easily navigate to footer messages in any application and have their screen reading software read the content.

## About Hearing Disabilities

People with hearing disabilities either cannot detect sound or may have difficulty distinguishing audio output from typical background noise. Because current user interfaces rely heavily on visual presentation, users with hearing disabilities rarely have serious problems interacting with visually oriented software. In fact, most users with hearing disabilities can use off-the-shelf computers and software. This situation may change as computers, telephones, and video become more integrated. As more systems are developed for multimedia, desktop videoconferencing, and telephone functions, designers will have to give greater consideration to the needs of users with hearing impairments.

Interfaces should not depend on the assumption that users can hear an auditory notice. In addition to users who are deaf, users sitting in airplanes, in noisy offices, or in public places where sound must be turned off also need the visual notification. Additionally, some users can only hear audible cues at certain frequencies or volumes. Volume and frequency of audio feedback should be easily configurable by the user.

Sounds unaccompanied by visual notification, such as a beep indicating that a print job is complete, are of no value to users with hearing impairments or others who are not using sound. Although such sounds can be valuable, designs should not assume that sounds will be heard. Sound should be redundant to other sources of information. On the other hand, for the aforementioned print example, it would be intrusive for most users to see a highly visible notification sign whenever a printout is ready. Visual notices can include changing an icon, posting a message in an information area, or providing a message window as appropriate.

Again, the key point here is to provide users with options and redundant information. Everybody using a system in a public area benefits from the option of choosing whether to see or hear notices. When appropriate, redundant visual and audio notification gives the information that is necessary to those who need it. If visual notification does not make sense as a redundant or default behavior, then it can be provided as an option.

Other issues to consider include the fact that many people who are born deaf learn American Sign Language as their first language, and English as their second language. For this reason, these users will have many of the same requirements for text information as any other user for whom English is a second language, making simple and clear labeling especially important. Similarly, as voice input becomes a more common method of interacting with systems, designers should remember that many deaf people have trouble speak-

Table 4. Assistive Technologies for Hearing Disabilities

| Assistive Technology | Function Provided |
|---|---|
| Telecommunications Device for the Deaf (TDD) | Provides a means for users to communicate over telephone lines using text terminals. |
| Closed Captioning | Provides text translation of spoken material on video media. Important computer applications include distance learning, CD-ROM, video teleconferencing, and other forms of interactive video. |
| ShowSounds | A proposed standard. Translates nonspeech audio (such as beeps) into screen flashes or other visual signals. Uses closed captioning technologies to provide text descriptions of speech from video and multimedia sources. |

ing distinctly, and may not be able to use voice input reliably. Like the other methods of input already discussed, speech should not be the only way of interacting with a system.

**Design Guidelines**

We have taken the design issues discussed in this chapter and condensed them into the list of guidelines shown in Table 5. This table also indicates which users are most likely to benefit from designs that follow the guidelines.

**Existing Keyboard Access Features**

Designers of Microsoft Windows, Macintosh, and X Windows applications should be aware of existing key mappings used by access features built into the Macintosh and X Windows (and optionally available for MS Windows). These features provide basic keyboard accessibility typically used by people with physical disabilities (see Table 2). In order to avoid conflicts with current and future access products, applications should avoid using the key mappings indicated in Table 6.

## INFRASTRUCTURE FOR ACCESS

We have identified a number of approaches to improve the accessibility of applications, but they leave many open issues. Even if all of these strategies

Table 5. Design Guidelines

| Design Guideline | Physical | RSI | Low Vision | Blind | Hearing |
|---|---|---|---|---|---|
| Provide keyboard access to all application features | X | X | X | X | |
| Use a logical tab order (left to right, top to bottom or as appropriate for locale) | X | | | X | |
| Follow key mapping guidelines for the local environment | X | X | X | X | |
| Avoid conflicts with keyboard accessibility features (see Table 6) | X | | | X | |
| Where possible, provide more than one method to perform keyboard tasks | X | X | | | |
| Where possible, provide both keyboard and mouse access to functions | X | X | X | X | |
| Avoid requiring long reaches on frequently performed keyboard operations for people using one hand. | X | X | | | |
| Avoid requiring repetitive use of chorded keypresses | X | X | | | |
| Avoid placing frequently used functions deep in a menu structure | X | X | X | X | |
| Do not hard code application colors | | | X | | |
| Do not hard code graphic attributes such as line, border, and shadow thickness | | | X | | |
| Do not hard code font sizes and styles | | | X | | |
| Provide descriptive names for all interface components and any object using graphics instead of text (e.g. palette item or icon) | | | | X | |
| Do not design interactions to depend upon the assumption that a user will hear audio information | | | | | X |
| Provide visual information that is redundant with audible information | | | | | X |
| Allow users to configure frequency and volume of audible cues | | | X | X | X |

were adopted for every current and future interface design, accessibility is ultimately limited by the capabilities of system infrastructure to support communication between assistive technologies and applications.

Recently, some progress has been made toward providing assistive access infrastructure. Parts of the OS/2 operating system, for example, were designed with blind access in mind (Emerson, Jameson, Pike, Schwerdtfeger, & Thatcher, 1992). The operating system infrastructure allows the IBM Screen Reader for Presentation Manager to obtain information about which window is receiving input, the contents of that window, and the layout of that window. For blind users, this infrastructure support means that they can obtain screen information which might be inaccessible on other systems (including font sizes

Table 6. Reserved Key Mappings

| Keyboard Mapping | Reserved For |
| --- | --- |
| 5 consecutive clicks of shift key | On/ Off for StickyKeys |
| Shift key held down 8 seconds | On/Off for SlowKeys and RepeatKeys |
| 6 consecutive clicks of control key | On/Off for screen reader numeric keypad |
| 6 consecutive clicks of alt key | Future Access use |

and styles as well as icon positions and labels).

In its most recent release, the X Window system has incorporated some similar access design features for users who are blind or have a physical disability (Edwards, Mynatt, & Rodriguez, 1993; Walker, Novak, Tumblin, & Vanderheiden, 1993). As a consequence, any developer of screen reading software can now develop X Window screen readers that can determine the state and content of windows. In addition, X Window software now provides support for keyboard enhancements such as StickyKeys (see Table 2).

Vanderheiden (1992b) proposed a standard cross-platform method for providing visual presentation of auditory information on computers. This strategy is based on a "ShowSounds" flag, which would be set by the user. The ShowSounds capability would allow users who are deaf, hearing impaired, or in quiet environments to request all information presented auditorially to be presented visually.[10] Multimedia applications, for example, would provide captioning for voice and alternative visual presentations for nonspeech audio. Such capabilities would benefit users learning a language, working in multilingual environments, as well as deaf users.

**Future Directions**

ShowSounds is an example of the system-level approach to providing access that is key to improving accessibility. Although current direct and assistive access capabilities are positive steps, much more can be done to improve accessibility. In the remainder of this section, we discuss some of the areas where system infrastructures must be developed to provide support for access.

Users with low vision, for example, typically view a relatively small por-

---

[10]As this book went to press, Microsoft announced that ShowSounds and other access features would be built into Windows 95.

tion of their display. Often they do not see color, and because of high magnification are unable to see new information or interface changes that are displayed off screen. In the future, users should have the capability to quickly navigate to regions where new information is posted. Future design efforts also need to expl ore how interfaces can gracefully handle a wider variety of viewing environments.

Note that some desktop environments (notably the Macintosh) do not provide comprehensive keyboard navigation capabilities allowing tab and arrow navigation through menus and controls. Assistive technologies requiring keyboard navigation in these environments have to provide that capability themselves. This approach increases the potential for incompatibilities with application key functions, and may not work effectively across all applications.

MS Windows, OpenWindows, and Motif all provide keyboard navigation, but none of these environments provides a standard method for navigating to read-only interface regions such as labels, message bars, and footer messages, nor do they provide a method for obtaining information about graphics. As a near-term solution, systems need to allow users to navigate to read-only text areas, and provide descriptive information for all nontext objects. Over the longer term, systems should be designed so they present and accept information in multiple modalities.

Blind users, for example, would benefit from both speech and nonspeech audio information based on video and graphics content. As in the case of closed captioning, such alternative presentations of information provide much more than access for people with certain disabilities. Audio descriptions of visual information would allow many users to access and manipulate graphical and video information over conventional telephones or while working on eyes-busy tasks.

Current interfaces are primarily visual, but as telephones become integrated with computers and multimedia becomes ubiquitous, access barriers for hearing impaired users will increase. Future systems should include support of video closed captioning. Closed captioning is useful not only for users with hearing disabilities, but also for users learning a second language and for a variety of contexts in which multiple languages are used. In addition, the descriptive text accompanying closed captioned video can be indexed to allow full-text searching of video and audio sources.

Looking farther into the future, adaptive interfaces along with intelligent agents can benefit users with disabilities by allowing them to interact with systems in the way best suited for each user (see Kuhme, 1993; Puerta, 1993). Unlike today's computers, future designs will allow most users to step up to any system and use it in a way best suited to their individual needs. We claim that an essential requirement for these future adaptive systems is that they account for the needs of users with disabilities.

# CONCLUSIONS

As with any other issue in human–computer interaction, the key to understanding the problem of accessibility is to understand the needs of users. We have suggested that users with disabilities are generally not considered in interface design. A shift from the narrow view of who constitutes "the user" to a broad view is the first step towards improving the accessibility of human-computer interaction for all users. Although, as we have suggested, much can be done by simply paying attention to the needs of users with disabilities, significant progress toward computer accessibility awaits improvements in the overall accessibility of the user environments.

Much is understood about the needs of users with disabilities that is not addressed by current systems. It is well understood, for example, that blind users need auditory and braille access to graphic information, that deaf users need visual access to audio information, and that users with physical disabilities need alternative means of interacting with their systems. To address these needs, future operating systems and user environments must be designed with accessibility in mind. To the extent possible, users should be able to "step up" to systems and use them without any adjunct assistive software or hardware. Where assistive software or hardware is required, system infrastructure should provide support for it to interoperate transparently with existing applications.

Multimedia and related future technologies hold both danger and promise for future prospects of accessible human–computer interaction. The danger is that without realizing the potential roadblocks presented by these emerging audio and video technologies, interface barriers far more daunting than those of today will be designed into future applications and environments. Many users could be locked out by interfaces that are usable only by people with a wide range of sensory and motor capabilities.

The promise is that this same mix of audio and video capabilities can significantly increase the accessibility of human-computer interaction. By providing the appropriate infrastructure to support multiple redundant input and presentation of information, systems of the future can allow users with disabilities to interact on their own terms while providing information in the form most useful to them. Not only will these users benefit, but ultimatley all users will benefit from interacting with systems that adapt to their needs and context.

# REFERENCES

Arons, B. (1992). Tools for building asynchronous servers to support speech and audio applications. In *5th Annual Symposium on User Interface Software and Technology* (71–78). New York: ACM Press.

Blake, T. (1985). *Introduction to principles and techniques for interface design.*

Tutorial notes for CHI'85 Tutorial.

Blattner, M. M., Glinert, E. P., Jorge, J. A., & Ormsby, G. R. (1992). Metawidgets: Towards a theory of multimodal interface design. In *Proceedings of COMPASAC 92* (pp 115–120). Los Alamitos: IEEE Press.

Brown, C. (1989). *Computer access in higher education for students with disabilities* (2nd Ed.) San Francisco: George Lithograph Company

Brown, C. (1992). Assistive technology computers and persons with disabilities. *Communications of the ACM, 35*(5), 36–45.

Casali, S. P., & Williges, R. C. (1990). Data bases of accommodative aids for computer users with disabilities. *Human Factors, 32*(4), 407–422.

Church, G., & Glennen, S. (1992). *The handbook of assistive technology*. San Diego: Singular Publishing Group

Edwards, A., Edwards, E., & Mynatt, E. (1993). Enabling technology for users with special needs. InterCHI '93 Tutorial. Unpublished manuscript.

Edwards, W. K., Mynatt, E. D., & Rodriguez, T. (1993, April).The Mercator Project: A nonvisual interface to the X Window System. *The X Resource*. Sebastopol, CA: O'Reilly and Associates.

Elkind, J. (1990). The incidence of disabilities in the United States. *Human Factors, 32*(4), 397–405.

Emerson, M., Jameson, D., Pike, G., Schwerdtfeger, R., & Thatcher, J. (1992). *Screen reader/PM*. Yorktown Heights, NY:IBM T. J. Watson Research Center.

Fels, D., Shein, G. F., Chignell, M. H., & Milner, M. (1992). Feedback control: Whose job is it anyway? In E. Gibson & G. Wright (Eds.), *Proceedings of the 25th Annual Conference of the Human Factors Association of Canada* (pp 81–84). Mississauga, Canada: Human Factors Society of Canada.

Furger, R. (1993). Danger at your fingertips. *PC World, 11*(5), 118.

Glinert, E. P., & York, B. W. (1992). Computers and people with disabilities. *Communications of the ACM, 35*(5), 32–35.

Griffith, D. (1990). Computer access for persons who are blind or visually impaired: Human factors issues. *Human Factors, 32*(4), 467–475.

Kuhme, T. (1993). A user-centered approach to adaptive interfaces. In W. D. Gray, W. E. Hefley, & D. Murray (Eds.), *Proceedings of the 1993 International Workshop on Intelligent User Interfaces* (pp 243–246). New York: ACM Press.

Lazzaro, J. J. (1993). *Adaptive technologies for learning and work environments*. Chicago: American Library Association.

Lewis, C., & Rieman, J. (1994). *Task-centered user interface design*. Electronic Shareware Publication.. This publication can be obtained via anonymous ftp from: ftp.cs.colorado.edu or email: clayton@cs.colorado.edu.

McCarthy, S. P. (1994). Treasury will equip disabled users. *Government Computer News, 13*(19), 111.

McCormick, J. A. (1994). *Computers and the American's with Disabilities Act: A manager's guide*. Blue Ridge Summit, PA: Windcrest.

McMillan, W. W. (1992). Computing for users with special needs and models of computer–human interaction. In P. Bauersfeld, J. Bennet, & G. Lynch (Eds.), *Conference Proceedings on Human Factors in Computing Systems, CHI `92* (pp. 143–148). Reading, MA: Addison Wesley.

Mynatt, E. (1994). Auditory presentation of graphical user interfaces. In G. Kramer (Ed.). *Auditory display: Sonification. audification. and auditory interfaces.* (pp.

533–555). Reading, MA: Addison-Wesley.

Newell, A.F., & Cairns, A. (1993, October). Designing for extraordinary users. *Ergonomics in Design.*, pp. 10–16.

Nielsen, J. (1993). *Usability engineering.* San Diego:Academic Press.

Perritt H. H., Jr. (1991). *Americans with Disabilities Act handbook* (2nd Ed.) New York: Wiley.

Puerta, A. R. (1993). The study of models of intelligent interfaces. In W. D. Gray, W. E. Hefley, & D. Murray (Eds.), *Proceedings of the 1993 International Workshop on Intelligent User Interfaces* (pp. 71-80). New York: ACM Press.

Sauter, S. L., Schleifer, L. M., & Knutson, S. J. (1991). Work posture, workstation design, and musculoskeletal discomfort in a VDT data entry task. *Human Factors, 33*(2), 407–422.

Schmandt, C. (1993). *Voice communications with computers: Conversational systems.* New York: Van Nostrand Reinhold.

Tognazzini, B. (1992). *Tog on interface.* Reading, MA: Addison-Wesley.

U. S. General Services Administration Information Resources Management Service (1991). *Managing information resources for accessibility.* Clearinghouse on Computer Accommodation.

Vanderheiden, G. C. (1983, October). *Curbcuts and computers: Providing access to computers and information systems for disabled individuals.* Keynote Speech at the Indiana Governor's Conference on the Handicapped, Indianapolis, IN.

Vanderheiden, G. C. (1990) Thirty-something million: Should they be exceptions? *Human Factors, 32*(4), 383–396.

Vanderheiden, G. C. (1991). *Accessible design of consumer products: Working draft 1.6.* Madison, WI: Trace Research and Development Center.

Vanderheiden, G. C. (1992a). *Making software more accessible for people with disabilities: Release 1.2.* Madison, WI: Trace Research and Development Center.

Vanderheiden, G. C. (1992b). *A standard approach for full visual annotation of auditorially presented information for users, including those who are deaf: ShowSounds.* Madison, WI: Trace Research & Development Center.

Walker, W. D., Novak, M. E., Tumblin, H. R., & Vanderheiden, G. C. (1993). Making the X Window System accessible to people with disabilities. In A. Nye (Ed.), *Proceedings of the 7th Annual X Technical Conference.* Sebastopol, CA: O'Reilly & Associates.

## APPENDIX

### Sources for More Information on Accessibility

Clearinghouse on Computer Accommodation (COCA) 18th & F Streets NW, Room 1213, Washington, DC 20405 (202) 501–4906

*A central clearinghouse of information on technology and accessibility. COCA documentation covers products, government resources, user requirements, legal requirements, and much more.*

Sensory Access Foundation 385 Sherman Avenue, Suite 2, Palo Alto, CA 94306 (415) 329–0430

*Nonprofit organization consults on application of technology "to increase options for visually and hearing impaired persons." Publishes newsletters on assistive technology.*

Trace Research and Development Center S-151 Waisman Center, 1500 Highland Avenue, Madison, WI 53528 (608) 262–6966

*A central source for the current information on assistive technologies as well as a major research and evaluation center. Trace distributes databases and papers on assistive technology and resources.*

### Conferences

ASSETS. Annual meeting of ACM's Special Interest Group on Computers and the Physically Handicapped. Every Fall. Location varies. Phone: (212) 626–0500

CSUN Conference on Technology and Persons with Disabilities. Every Spring in Los Angeles, CA. Phone: (818) 885–2578

Closing the Gap Conference on Microcomputer Technology in Special Education and Rehabilitation. Every Fall in Minneapolis, MN. Phone: (612) 248–3294

RESNA. Conference on rehabilitation and assistive technologies. Every Summer. Location varies. Phone: (703) 524-6686

# Chapter 5
# Pen-Based Interaction for a Shared Telephone Workspace

Andrew Hunter
*Hewlett Packard*

## INTRODUCTION

General-purpose computers still exist as a distinct product category; however, computer technology is now cheap enough to be embedded in a wide variety of dedicated devices. As the boundaries between computers and noncomputers disappear, the fields of computer interface design and product industrial design are beginning to merge.

Computers are also evolving to satisfy a wider range of personal and business needs. Recent research has demonstrated improvements in business communication by the use of shared computer workspaces (Bly, 1988; Whittaker, Geelhoed, & Robinson, 1993). So far, most research in this field has focused on the use of general-purpose computers that are connected to each other over data networks.

At Hewlett Packard Laboratories we have begun experimenting with some new ideas for dedicated computers and their user interfaces. The device described here is a pen-based device called DeskSlate (see Figure 1). DeskSlate is a prototype interactive fax machine. It can send documents to conventional fax machines, and receive documents from them, but when a call is made from one DeskSlate to another, the faxes can be viewed, discussed and annotated simultaneously by both parties. The fax transmissions between the machines are mixed on the same ordinary telephone line as the ongoing verbal discussion. DeskSlate thus provides the benefits of a shared workspace without the need for a computer network.

Figure 1. A DeskSlate interactive fax prototype.

The purpose of this chapter is to convey some of our experiences in design-ing the user interface of the DeskSlate prototype. In particular, it describes how the design was shaped by the technologies of pen-based interaction and flat panel displays, as well as by the more obvious influences of the collaborative task it supports. We hope that the account of our experiences will be of value to anyone developing systems that use these technologies or support collaboration.

This chapter does not attempt to provide a complete description of the func-tions of the finished DeskSlate prototype, or to describe all the issues that arose during its development. Instead, it focuses on the design issues that are novel or most likely to be relevant to other projects. The issues are conveyed through examples of design elements that have evolved as our understanding of the issues increased.

We begin with a brief description of the overall process that allowed us to discover and respond to design issues. In the following sections we then describe the specific issues raised by pen-based interaction and flat panel dis-play technologies, and discuss their impact on the DeskSlate interface. We also review the issues raised by the collaborative nature of a shared telephone work-space. Throughout the development of Desk3late, one lesson has dominated over all others: The production of a design must not be divorced from the details of its eventual implementation, especially when it exploits new technologies.

## THE INTERFACE DESIGN PROCESS

Superficially, the result of a successful interface design process is a tasteful outward appearance for device functions, but of course there is more to a design than its outward appearance. To ensure that the functions of the DeskStlate prototypes were appropriate and easy to operate, as well as aesthetically pleasing, our design process also included regular ergonomic evaluations of the proposed designs. However, even the combined constraints of design aesthetics and interface ergonomics were not sufficient to ensure that the device would serve its users well.

Designs must also take account of the means by which they are implemented (Lambert, 1993). In the case of DeskSlate, some of the underlying technologies were relatively new. For example, we had to identify and work within the constraints imposed by flat panel displays and pen sensors. These technology limitations were perhaps as significant in shaping the DeskSlate prototype as the human limitations that were identified during our ergonomic evaluations.

At the outset of the DeskSlate project, the constituent technologies and precise device functions had not been chosen; therefore, many of the design constraints were also unknown. In the absence of a firm specification, our design process was established as a strategy for exploring the details of both the problem and its solution.

### Exploratory Design

The design process was iterative (see Figure 2). In the top half of the cycle, ideas were picked from within the design space, and were embodied in some testable form. In the bottom half, the embodiments were evaluated, and further constraints were defined to exclude any of the identified faults from the design space. Thus the problem definition and the solution both took shape in parallel.

The diagram represents a flow of information rather than a strict sequence of activities. In particular, we did not begin at a single point in the cycle. At the outset, we had both a product idea and a problem definition based on previous evaluations of shared workspaces (Whittaker, Geelhoed, & Robinson, 1993). The process had wheels within wheels. Although the main concept of the solution may only have gone round the cycle two or three times, some of the design details iterated very rapidly.

Design limitations due to intended users, environments, tasks, production

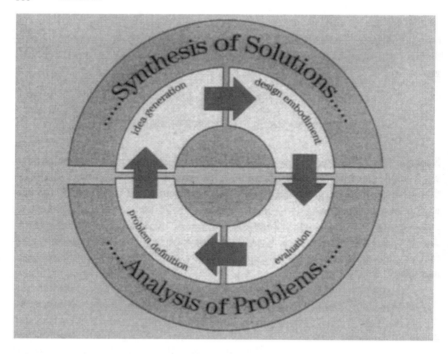

Figure 2. The DeskSlate interface design cycle.

methods, and sales channels all exist whether or not they have been stated in a design brief. If, as in this case, they have not been stated, they must be discovered during the design process. Our initial problem definition required considerable refinement to encompass relevant design constraints, but many of those constraints were hard to identify without some degree of trial and error. We thus generated tentative solutions from early in the design process (Darke, 1979). The early solutions were disposable—they were chosen to challenge and refine our problem definition rather than to form part of the final design. Later, as the problem became clearer, the broad concepts of a solution were retained, and the finer details were explored in the same manner. Eventually even they were retained. The process thus shifted gradually from an analysis of the initial problem to an evaluation of the final solution.

**Design Embodiment**

Most of our interface designs began as sketches on paper. This allowed us to specify the ideas rapidly, and discuss them within the team before committing the designs to code. In fact, most of our ideas were never prototyped. A vast number of sketches were rejected for reasons such as anticipated performance problems, unsatisfactory aesthetics, and unclear methods of operation. The process of idea generation and evaluation was informal and rapid. Its purpose was not to prove conclusively that any one design was better than another, but to select the most promising candidates for further prototyping.

Sketching continued throughout the development process, even after most of the device functions had been embodied in software. The sketches were so quickly produced that we were able to explore a wide range of alternatives to the solutions which had already been implemented. Occasionally, we made fairly radical changes in order to test promising new ideas. Without this willingness to consider changes, the impact of each successive design refinement would have quickly diminished. We thus sometimes had to undo successful elements of the design in order to open new opportunities for further improvements.

An obvious limitation of sketched designs is that they cannot specify the active or interactive aspects of an idea. However, a static sketch of an interface should be able to convey an impression of those behaviors. If it doesn't, it may be the design that is at fault, resulting in a user of the implemented design being equally unable to anticipate interface behaviors.

Unfortunately, ideas that work well on paper don't always look good on a computer screen. To assess the most promising sketches, we therefore used a simple image editor to produce screen mock-ups. The mock-ups took much longer to produce than the original sketches. Consequently, they were produced in far smaller numbers, usually to test specific visual effects. By producing mock-ups, we found that some of our ideas required a higher resolution or greater tonal depth than was available on the prototype display. We also discovered that the visual quality of the images varied from one display type to another. However, by ensuring that we viewed the mock-ups on the chosen hardware we could see how the screens would look during use, and could even factor display peculiarities into the design process.

The activities of interface design and implementation are often quite separate. Designers consider user needs and implementers consider interface technologies. However, separation is not always beneficial. Design and implementation should be very closely linked. Whether a product is constructed in wood, steel, or software, its final form must fit the material and production methods as well as its intended function. Architects need to understand the structural properties of bricks and mortar. Fashion designers need to understand the properties of different cloths and stitches. Although interface designers do not need to implement their own designs, they should take account of the opportunities and limitations of the technology.

Like the ergonomic aspects of the DeskSlate design, the technical aspects also benefited from empirical testing. Working DeskSlate prototypes were built from early in the development process to allow us to identify and overcome technical problems. We studied the technical limitations and interactive performance of the software running on the final hardware, and tuned our designs accordingly. We began with basic implementations of the core functions, then refined and extended the functionality, taking design ideas from the screen mock-ups. At regular intervals we released versions of the software for empiri-cal evaluation and for demonstration purposes.

## Evaluation

Comparative studies of voice compression schemes, and of pen characteristics, were carried out using rank-order evaluation techniques. However, most of the software evaluations were to assess the quality of a single evolving prototype. In an ideal world, there would be time to build several alternative solutions to each interface problem for comparative evaluation. In reality, there wasn't even enough time to solve all the known problems.

Even the simplest of experiments can take a considerable length of time to conduct and analyze, so the DeskSlate evaluations had to be chosen carefully to focus on the most critical interface issues. Furthermore, the evaluations had to be scheduled to occur when stable prototypes became available, and to produce results in time to influence the following development stages. Early on, while the DeskSlate prototypes were fairly primitive, the evaluations sought to discover fundamental problems with telephone collaboration and hence provide valu-able and timely design insights. Later, they sought to discover remaining prob-lems with the prototypes themselves. The emphasis of the evaluations thus shifted from desirable applications of the technology towards details of its oper-ation.

Evaluation does not only serve to identify design flaws. A design could be improved indefinitely, so evaluation can be used to identify the point at which further improvements are unnecessary. Evaluation can also be used to identify design elements that should not be changed. This is particularly relevant because redesign can rarely be limited to the elements that are flawed due to the need for design coherence across device features.

Laboratory experiments were designed to test specific user-interface fea-tures and overall interface complexity. They focused mainly on the operation of the device, ensuring that it functioned as we had planned. However, from the early stages of the DeskSlate project, we also provided a prototype link to be used for real business communications between the Laboratories in England and a Hewlett Packard site in France. Naturalistic observations of this link were not repeatable as under the controlled conditions of laboratory experiments, but they covered a wider range of collaboration tasks and allowed a more realistic assessment of the benefits to be gained. Finally, the DeskSlate field trials allowed us to test our theories about collaboration over much longer periods of use. They also allowed us to evaluate how well the devices worked in a variety of office and home environments.

In the field trials, we were more interested in general lessons about appro-priate DeskSlate functions than in specific interface details. However, a consid-erable level of design refinement was necessary to ensure that a poor interface to the functions did not taint the results. In addition to the core functions to be evaluated, such as collaborative document annotation, the prototypes had to provide essential support functions, such as those for storing or printing the documents. Furthermore, the prototype software and hardware had to be reliable enough to function for an extended period without technical support.

Despite our efforts, the quality of the prototypes significantly hampered the trials. Like products, field trial systems must be designed to fit in with existing work environments and processes. The DeskSlates had been intended to satisfy users' needs for fax transmission and receipt, but the early prototypes did not display documents with sufficient visible detail. Consequently, most of the participants continued to use their existing fax machines. They used their DeskSlates when they wanted to discuss documents with other field trial participants, but we had not provided a satisfactory method for scanning their faxes or other paper documents into the prototypes. Despite these problems, we learned many useful lessons about DeskSlate functionality, and found some unexpected benefits. For example, the ability to retain written or sketched records of telephone discussions provided long-term benefits that had not been observed in the earlier experiments (O'Conaill, Geelhoed, & Toft, 1994). We also found that the faxing capabilities were used as a form of quick, handwritten electronic mail.

## DESIGN INFLUENCES OF PENS

Pen-sensitive flat-panel displays provide some exciting new options for device interfaces; however, they are sufficiently different from previous forms of input device to require new design approaches. The opportunities and constraints of pen technology had a huge impact on the evolution of the DeskSlate interface.

### Conventional Manual Interfaces

Most conventional items of consumer electronics, such as hi-fi systems and video recorders, are operated via arrays of physical controls. Whether the knobs, buttons, sliders, and switches are on the face of a device, behind a hinged or removable panel, or even on a separate remote-control unit, their number and arrangements are fixed in hardware. Usually, each control serves a specific purpose.

Some of the simplest and most elegant controls, such as volume knobs, serve to display an aspect of device state as well as to accept user input to change it. However, these manipulable displays are becoming less common as devices become more sophisticated. Consider a volume knob that allows a user to view and set the volume level of a device. If the volume represented by the position of the knob can automatically change to a pre-set level, or be changed with a remote control or other means, the knob must be able to physically rotate itself to display the new state. To avoid the complexity of this mechanical feedback, many device manufacturers are now using simpler, stateless controls such as volume-up and volume-down buttons, with visual feedback elsewhere on the device.

Furthermore, with increasing functionality on ever-smaller devices some of these stateless controls are now being used for multiple purposes. For example, a single pair of buttons marked only with a plus and a minus are sometimes used to control an aspect of a device that has been selected from a menu of options on a display panel. The level of internal sophistication and the sheer number of device features are therefore forcing a change away from more traditional dedicated controls. Knobs and switches designed for specific functions are being replaced by generic buttons that do not convey any device state. Optimization is being sacrificed for flexibility.

Although they cannot rival physical controls for tactile feedback, pen-sensitive display panels can address some of these issues. Instead of fixed arrangements of knobs and buttons for detecting user input, a whole panel is able to detect input. Because it is also a display, software controls can be arranged freely on its surface, and can be rearranged to suit specific phases of interaction. Like volume knobs, the controls may serve a display purpose, but they do not require a complex mechanical feedback mechanism. Their visible states may change equally easily, whether to reflect user actions on them, or to reflect changes from within.

## Mouse-Operated Interfaces

The knobs and thumbwheels of hardware devices are designed for manual input and therefore may be inappropriate for pen-operated devices like DeskSlate. Where, then, can we look for inspiration when designing pen interfaces? Since the advent of graphical input devices such as mice, computer interfaces have evolved to take advantage of similar freedoms from fixed arrangements of hardware controls. Some portable computers now even have pen-sensitive displays, so there is little doubt that pens can replace previous graphical pointing devices, but is it appropriate for them to do so?

Current computer interfaces may work well for mouse input, but pointing devices are not all alike (Buxton, 1986) so the interfaces cannot be assumed to work equally well for pen input. In particular, some common interface controls, such as menus and scroll bars, rely on feedback that may be obscured by a hand holding a pen.

Mice and pens allow users to press, turn, and slide controls as for manually operated physical buttons, knobs, and sliders. In addition, they allow users to annotate and rearrange things that are displayed. There is no doubt that a pen is superior to a mouse for written and sketched annotations; however, there is experimental evidence to suggest that dragging a displayed object is significantly more error prone with a pen than with a mouse (MacKenzie, Sellen, & Buxton, 1991).

A computer can track and respond to the position of a mouse as it moves over a table top whether or not a mouse button is being pressed. Pressing a pen against a pen-sensitive display is treated as equivalent to pressing the button of a mouse; however, when the pen is not pressed against the display, it is free to move onto and off of the surface as well as across it. Consequently, it is not always possible for a computer to track and respond to the position of a pen as it would have done for a mouse.

This difference can also be seen clearly when using the devices to double click on a displayed item. A mouse remains steady on a surface while its button is double clicked, but a pen tends to be lifted off the surface during the corresponding double tap, making it hard to aim the second tap at precisely the same point.

Further problems arise from the direct application of the pen to the screen. A mouse controls the direction and approximate speed of a motion on the screen. When dragging an object with a mouse, small discrepancies between the speed of the mouse and the movement of the thing being dragged do not matter much. With a pen, the intention is to affect something at the actual point of contact with the screen. Without a very powerful processor, however, an object being dragged will lag visibly behind the pen.

More studies are necessary to determine the range of input actions that are well-suited to pen-based interfaces. In general, special care must be taken when designing and arranging controls for pens. If interfaces are redesigned, however, there are other properties of pens that should be exploited. Because they are applied directly to objects on a screen, pens give a very powerful feeling of direct control. To enhance this feeling, we have designed the DeskSlate interface around very physical manipulations like pressing buttons and writing on documents. More specifically, we have arranged controls to avoid having the users' hands obscure feedback.

## DeskSlate's Software Surfaces

Nearly all modern products have an external case that conceals and protects the internal workings. The case also serves as a formal separation of the components that are intended to be manipulated by a user, from those which implement the internal behavior. It is the responsibility of a product designer to express device state and behavior through externally visible surfaces, and to provide appropriate controls to affect the behavior.

Designers have not always separated relevant and irrelevant state, or manipulable controls and nonmanipulable workings. These distinctions have evolved over the last century to provide significant improvements in device interaction. The lessons are also applicable to software interfaces: We have designed the DeskSlate interface as a set of surfaces that convey relevant state and have pen-operable controls to trigger internal behaviors.

In the real world, identical actions with a pen can have different effects depending on the materials to which they are applied. If a pen is drawn across paper it will leave a trail of ink. If it is drawn across glass it will probably skid over the surface without leaving a mark. If it is applied to a soft surface it may grip and drag the surface instead of moving across it. In all cases the action is the same, but the result is determined by the context in which the action is performed. In DeskSlate we make use of the same technique to increase the range of effects possible by a small number of simple pen actions.

In many interfaces, a single action can be interpreted in multiple ways. In the case of the generic plus and minus buttons that were discussed earlier, the user had to make a selection from a menu to assign a meaning to the buttons before operating them. Computer menus are often used in the same way, to assign a specific interpretation to mouse actions before they are performed. In DeskSlate, we have tried to ensure that all manipulations are distinguished by the surfaces to which they are applied rather than by preliminary actions to instruct the device to assign the desired interpretations.

In DeskSlate, the fundamental actions are to press and to draw the pen over the screen. No special interpretation is applied to any other pen actions. In particular, double clicking has no significance other than two short presses. The primitive actions were chosen to be easy to perform with a pen and also to resemble physical manipulations. The manipulations should seem to cause the corresponding surface effects, not just invoke them by some unseen process. The primitive actions and surface responses constitute a sort of naive physics that applies throughout the interface and helps users to predict and explore behaviors (Owen, 1986).

As an illustration of the concept of software surfaces, imagine an electronic version of a diary page (see Figure 3). The interface consists of five distinct surfaces—an enclosing surface with an aperture in the center of it, a diary page visible through the aperture, and three buttons. Activities planned for the next hour or so can be seen on the visible portion of the diary page. As time passes, the page automatically moves so that the current time aligns with arrows on the enclosure. Two buttons allow the user to scroll the page up and down in order to add or view appointments elsewhere on the page. A third button returns the page to its normal position, with the current and impending appointments visible.

Because they are distinct surfaces, the enclosure, page, and buttons can interpret pen actions differently. Pressing the pen on the enclosing surface has no effect; pressing the pen on the page makes a mark on the page; pressing the pen on one of the buttons causes the button to sink momentarily into the case.

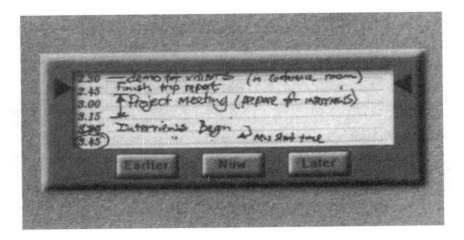

Figure 3. Mock-up of Diary Software

and then as an indirect result invokes the corresponding internal function to move the diary page. Note that the diary contains no surfaces that can be both marked and dragged—this is a general design constraint imposed to simplify our interaction model.

### Visibility of Surface Behavior

The whole of a DeskSlate display is composed of surfaces. Any manipulation can be applied freely to any one of the surfaces. For example, there is nothing in the diary interface to stop a user trying to write on the unmarkable surface surrounding the diary page. If the user tries to write on it, there is no need for the interface to display an error message, because the failure of the pen to make a mark is a sufficient explanation of the problem. However, it is an indication of poor design of a surface appearance if a user ever expects an unmarkable surface to be markable, or an undraggable surface to be draggable, or any surface to behave differently than the way it does.

Surfaces are not only important in the interpretation of users' actions, they are also important in users' understanding of the interface behaviors. The situation is the same when users look at real-world devices such as hi-fi systems. They must visually distinguish each of the control surfaces, and must anticipate whether they are intended to be pressed or rotated or manipulated in some other way. Surfaces are said to *afford* particular manipulations (Gibson, 1977). Well-designed surfaces convey appropriate affordances and hence avoid inappropriate manipulations. For example, well-designed buttons look pressable, well-designed knobs look rotatable, and so on. It is thus the responsibility of the designer of the diary to ensure that its visual appearance leads users to perform appropriate manipulations of each its constituent surfaces.

Affordances work equally well at guiding both first-time and expert users. Other techniques for guiding interaction, such as explicit on-screen instructions, would help novices but might even hinder experts. Unlike explicit instructions, affordances provide subconscious visual cues to appropriate interactions. In most cases, the surfaces of the DeskSlate interface seem to have conveyed the appropriate affordances. Users have not expected to be able to annotate buttons, or drag them, or do anything other than press them. They have not expected to be able to annotate the screen background, except on one occasion when we temporarily replaced the textured background with a uniform pale gray tone. So it seemed that the texture successfully conveyed our intention that the background could not be annotated, whereas the uniform tone afforded annotations.

**Indirect Effects of Surface Manipulations**

The DeskSlate prototype has been designed to ensure that all manipulations of surfaces have an immediate visible effect or no effect at all. We judge effects to be "immediate" if they seem to result directly from the action of the pen. One source suggested that a delay of more than a tenth of a second would destroy this illusion of direct manipulation (Nielsen, 1993). Thus, if a user writes on a surface the marks will either appear immediately as if due to ink flowing from the tip of the pen, or the surface is not markable. Similarly, if the user tries to drag a surface, the surface will move immediately as if gripped by the pen, or the surface is not draggable.

If for any reason we were unable to guarantee the immediacy of an effect, we designed the surfaces of DeskSlate to prevent it. For example, if a surface was too large for us to guarantee that it could be displayed fast enough to seem to be under the control of the user's pen, we did not allow it to be dragged. Instead, if it was necessary for the surface to be moved to a new position, we provided an indirect mechanism by which the movement could be achieved. Thus, pressing a button would have an immediately visible effect on the button, but would also trigger an automatic movement of the surface. Such indirect control is sometimes preferable even when a direct manipulation with the pen could achieve the same result. For example, we could have designed the diary interface to enable the page to be repositioned by a direct action of the user. However, it is easier for the user to press a button to automatically bring the impending appointments into view than to manually drag the page to the same position. Notice that although the surface around the diary page serves no interactive purpose, it does seem to enclose the mechanism that moves the page. We believe that it is important to associate internal behaviors with specific displayed objects in order to provide an understandable relationship between the actions, behaviors, and effects.

In DeskSlate, the only way a user can trigger an internal behavior of the prototype is by manipulating a visible surface. If the manipulation has any effect on the surface it will be immediately visible, even if the effect of the triggered behavior is less immediate. In most cases, the behavior introduces a noticeable delay. For example, when a DeskSlate user presses a button to fax a document, the transmission cannot be instantaneous. Similarly, when a user presses a button to bring a document onto the screen for editing, it takes a second or two for the document to appear. When an indirect effect is subject to delay, we try to give intermediate feedback in the interface to show that the effect is pending, and to show progress towards its achievement. Again, the feedback is related to the surfaces that seem to enclose the behavior, rather than at an arbitrary location.

If surfaces are designed to convey the direct effects that they afford, users may be able to achieve some goals very simply. For example, when a user wishes to make a note of an appointment, and the markable surfaces are obvious, the user only has to choose one on which to write the appointment. However, when a user's goal cannot be achieved by a direct manipulation, he or she must search the interface for the appropriate indirect mechanism. Thus the surfaces must also be designed to convey the indirect effects which may result from any triggered behaviors. It is important that a user can quickly decide whether or not an appropriate indirect mechanism exists; in other words, a major role of an interface is to convey its own limitations.

Users do not plan interactions in isolation from the interface: They view and respond to it. Sometimes they will have goals in mind without knowing exactly how they intend to achieve them. If the interface is well-designed, when the appropriate situations arise, the users may spot ways to achieve the goals. These are *situated actions* (Suchman, 1987). On occasion, new goals may even be suggested by the visible features of the interface. For example, a user may spot a button to print a document, and may decide to do so even though it had not been a prior intention. Such opportunistic interactions are enabled by what we call *feedforward* mechanisms of which visual affordances are an example. In general, feedforward mechanisms are cues to guide or motivate future interactions.

## Controls Designed for Pens

DeskSlate's controls were arranged carefully to avoid having the user's hand obscure relevant feedback. They were designed to be easy to operate with a pen either by pressing or by simple movements of the pen over the screen. The controls were also designed to convey their method of operation. For example, controls that were operated by moving the pen over the screen had appearances to convey whether their motion was constrained to a straight line, or to rotation about a point (see Figure 4).

The motion of a pen over a screen does not provide any tactile feedback of the effect on the surface underneath. In particular, if the surface cannot be dis-

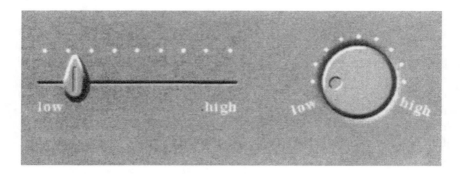

Figure 4. Controls which convey linear or rotational movement.

played quickly enough to seem to move at the same speed as the pen, the movement of the pen cannot be forcibly slowed to the speed of the surface. To preserve the feeling of direct manipulation, it is important that any movement of a control seems to result from physical contact with the pen. Scroll bars and controls based on hardware knobs and sliders are not ideal for operation with a pen because contact cannot be maintained between the pen and the point being dragged.

A solution to this problem is to design moveable control surfaces with one or more apparent hollows. To move a surface, a pen must be kept within a hollow. If the pen moves too fast, it will be seen to move out of a hollow and the manipulation will cease. Consider the constrained motions of a slider and a rotatable knob. In these cases, the movement of the pen while in contact with the surfaces should be constrained to a straight line and an exact circle, respectively, but of course it cannot be constrained. Thus, allowing the pen to move within a hollow means that the track of the pen and the track of the original point of contact on the surface can differ by an amount determined by the size of the hollow (see Figures 5a and 5b).

**DeskSlate's Buttons**

The DeskSlate interface has been built for direct manipulation with a pen, yet a large number of its functions are button-operated. Although the buttons are directly manipulated, the functions that they control are not.

When a desired change to the state of the system can be achieved by a simple representative action, a direct manipulation is probably an appropriate means to achieve the change; however, an interface cannot provide a direct manipulation to achieve every desired effect. For example, when a user wishes to print a document, the visualizable goal is to have a copy of the document sitting in the output tray of the printer. There is no manipulation of the displayed

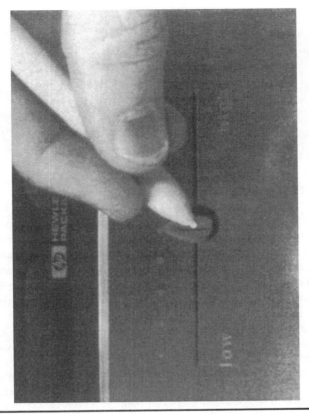

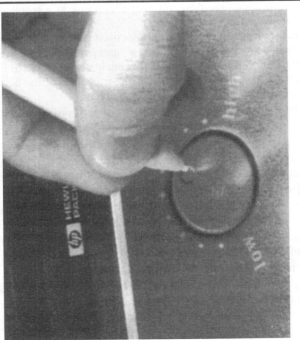

Figures 5a and 5b. Hollows allow small amount of movement between the pen and controls.

Figure 6. Early rectangular buttons.

document that can achieve this result. A common solution to this problem is to display a representation of the printer on the screen so that the user can drag the document onto it. Dragging the document is certainly a direct manipulation, but dropping it on the displayed printer was not the goal. The goal was for the real printer to print a copy of the document. The user thus has to learn that the action of dropping the displayed document on the displayed printer initiates a process which results in a printed copy on the real printer.

If a user has to press a button to print a document, the action is less metaphorical, but no less direct. The purpose of an interface metaphor is to relate a device function, and perhaps also a method of operation, to something a user has already experienced. We believe that in this case, an appropriately labeled button can express the function and operation more simply. There is also evidence to suggest that users can point at a target such as a button faster and more accurately than they can drag something to a target of the same size (MacKenzie, Sellen, & Buxton, 1991).

Because buttons initiate internal functions of a device, they do not restrict the influence of the functions to a single manipulable item—a single press of a print button can thus cause multiple documents to be printed. Regrettably, this also requires a mechanism by which the user can select the document or documents to which the initiated process will apply.

The buttons used in the earliest working prototypes were of the type standardized in the Open Software Foundation's Motif graphical user interface (Kobara, 1991). They were rectangular with beveled edges, giving a three-dimensional appearance (see Figure 6). Because of their rectangular shape, these buttons were compatible with standard procedures for detecting pen actions. However, individual buttons were not very distinctive. In laboratory tests, even though the early prototype had very few functions some users had difficulty finding the buttons for the functions they required. In other cases, users pressed the wrong buttons even after they had correctly described the function of each. It seems that after a short time, users select buttons according to their general appearance and position rather than by inspecting their labels.

Later, we tried some alternative appearances for rectangular buttons. We were able to give the buttons a rounder look that suited the industrial design of the hardware better (see Figure 7). We also found that we could vary the surface tones and textures to distinguish individual buttons more clearly.

Encouraged by these improvements, we were keen to experiment with non-rectangular buttons. The principles of surface interaction do not distinguish buttons as a special type of surface. Some surfaces can be moved with a pen, and others can be annotated. Buttons are just surfaces which happen to respond to being pressed with the pen—there are no rules which dictate that pressable surfaces must be rectangular. For these reasons, we decided to develop our own procedure for checking whether a pen had been pressed over the display of an arbitrarily shaped surface.

Our experiments with the new buttons also taught us some important lessons about button affordances. Like most features on computer interfaces, the previous buttons had black outlines. We removed these outlines from the new buttons because their edges were defined adequately by highlights and shadows (see Figure 8). The effect was surprising. Without outlines the buttons no longer looked pressable; instead, they looked like raised bumps on a continuous solid surface.

Consistency within and across interfaces is often quoted as a significant contribution to ease of use (Nielsen, 1993). However, having seen the considerable impact of such a simple design feature as a button's outline, we decided to explore the deliberate visual expression of pressability as an alternative to consistency. If we could not design buttons that looked inherently pressable, we could fall back on consistency with other computer interface buttons as a means to convey their behavior. Buttons were one of the design elements that passed very rapidly around the design cycle. On each of the many iterations, we used sketches, mock-ups or working prototypes to assess their affordances.

The outlines around early buttons had taught us the first rule of pressable surfaces, which is that they must look like separate surfaces that can move independently. In accordance with this rule, we produced an improved button design

Figure 7. A variation of the rectangular button design.

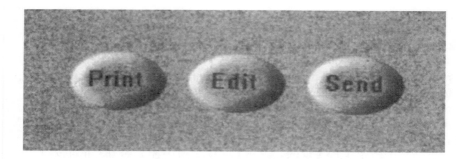

Figure 8. A button without an outline looks less pressable.

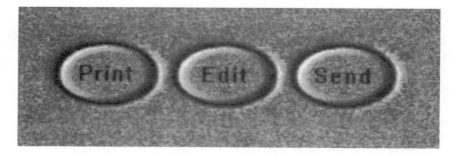

Figure 9. A later button design with clearer affordances.

Figure 10. The buttons in the current Deskslate prototype.

that emphasised the physical separation of the button and the surrounding surface (see Figure 9). Furthermore, the surrounding surface was made to seem to
stop at the edge of the button instead of passing underneath it. This makes it
easier to imagine pressing the button down into the surface than to imagine sliding it over the surface. The second and third rules thus became to suggest the
freedom to move into the screen, and to suggest constraints preventing lateral
movement across it.

We did not set out to find the ultimate button design. The design that
remains in the prototype to this day is rather old-fashioned (see Figure 10). It

was chosen because it looked more pressable than any other we tried and had the most obvious visual feedback of being pressed. Conspicuous visual feedback was important because we hadn't implemented an audio feedback mechanism, and of course the pen and screen gave no tactile confirmation of button operation. Although the affordances of this design seem to be strong, they are not entirely appropriate. The large and slightly concave top seemed to afford pressure with a finger rather than a pen—this was a surprisingly common mistake for novice users.

Although the manipulable surface of a button conveys its method of operation, there is normally also a button label to allow the user to predict the indirect effect of its operation. We would have liked to have found some appropriate graphic symbols to convey the button functions, but we have resorted to placing explicit text on the button tops. Textual labels suffer from language dependency, and they demand relatively large buttons or plenty of space nearby. However, graphic symbols are often ambiguous and may not provide any significant advantage over text (Benbasat & Todd, 1993). These visual problems are not specific to software product design so we will not discuss them at length here. However, the best solutions for a software product may not be the same as for a hardware product design. The design constraints and production issues are very different. Operation with a pen also has a significant impact. For example, a pen doesn't hide the button top like a finger would. Thus, the best place to display indirect feedback, such as a flashing LED, might be on the surface of the button itself.

## DeskSlate's Control Panels

Originally, we believed that a small set of permanently visible buttons would be able to provide all the necessary functions of DeskSlate. In an early prototype, the buttons were arranged in a column down one side of the screen, using space that would not be obscured when a fax page was being viewed. We did not like the buttons in this position. We were concerned about the implications for left- and right-handed users. Would the software need to be reconfigured to suit its user? We decided to arrange the buttons more symmetrically across the bottom of the screen, but that introduced another problem—a fax page was going to be approximately twice the height of the screen, and we wanted it to look like a whole page that extended off either the top or bottom of the screen depending on the portion being viewed. How could we avoid obscuring the buttons?

The solution was to remove the buttons from the screen background, and arrange them on a separate horizontal bar (see Figure 11). As a page was brought onto the screen, it would appear to slide under the bar. The page could be scrolled up or down without hiding the buttons. The bar had the added benefits of grouping the buttons in a single location and reducing the clutter on the screen background.

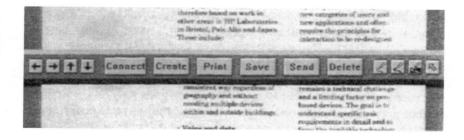

Figure 11. Buttons arranged on a horizontal bar.

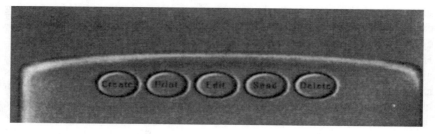

Figure 12. Buttons arranged on a sliding panel.

We were very careful not to add unnecessary functions to the prototype, but as development continued we found that the list of essential functions grew quite long. Our tests had also shown that users were confused by the continuous presence of buttons that were not always relevant. We were thus forced to drop the idea of an unchanging set of buttons. We divided the buttons into two smaller sets, and replaced the single button bar with two panels that would slide into view from the bottom of the screen when they were needed (see Figure 12).

The interface is now moded, but only in the sense that the two sets of controls can move in and out of view. None of the surfaces change their behaviors. Because goals are achieved by manipulation of what is visible, the modes would be a problem if actions were frequently required to switch the hidden set of controls back into view. However, users' goals seem to shift between those related to the details of document content, and those related to whole pages. We have grouped the controls to fit these interaction trends. One panel automatically appears when a document is selected for shared discussion or private review. It provides viewing and editing controls. The other panel appears when the stored documents are being surveyed. It provides controls such as to print, fax, or delete sets of pages.

The button panels did not result from any great interface plan. They evolved as a result of regular testing and fairly pragmatic design revision. Tests of the new designs confirmed that the panels overcame the earlier interaction problems. The buttons available on each panel were easy to see at a glance. The changes in available buttons were conspicuous without being disturbing, and they seemed to match natural modes of the interaction. In tests, users rarely looked for functions without bringing the correct button panel into view first. Controlling the panels seemed to become an accepted part of the tasks.

## DESIGN INFLUENCES OF THE DISPLAY AND PEN SENSOR

DeskSlate is basically an enhanced fax machine. It allows users to preview faxes on a screen before printing or deleting them. It can also function interactively, allowing users to view a document across a telephone line while it is being annotated. In both of these roles, the DeskSlate functionality relies on the legibility of documents on a flat panel display. Even though it provides capabilities that are not possible with paper, the visual quality of displayed faxes must be as close to the printed equivalent as possible.

For most current pen-operated devices, annotations need only be legible to the person who wrote them, or be consistent enough for an automatic handwriting recognition function. In DeskSlate, the pen is for communication with another person, so the quality of annotations is much more important. The annotations must be smooth and legible on the screen and must also look good when documents are faxed from the device, especially if they are to be sent to people who are not familiar colleagues.

### Display Resolution and Grayscale

Flat panel image display technologies are still relatively new. Most of the displays suitable for our application were developed for portable computers. They all had 640 x 480 picture elements (pixels) and a resolution of around 80 pixels per inch. Most of them could display 16 levels of grayscale.

Resolution and grayscale have a significant influence on page viewing and handwriting quality (Cushman & Miller, 1988). DeskSlate documents are displayed close to life size so that the full width of a fax page can fit across the display. In early prototypes, the pages were displayed without grayscale and as a result it was difficult to read small text on some faxes. Grayscale is now used to improve the perceived resolution. A fax page has 200 pots per inch so each group of three by three dots on the original is displayed as a single dot of black, white or intermediate gray, depending on the numbers of black and white dots in the group. As a result of this change, documents are now much easier to read; however, we have still not taken advantage of grayscale for improving the legibility of annotations. Annotations are laid down as lines of black dots. The individual dots are very noticeable especially on small handwriting, which tends to

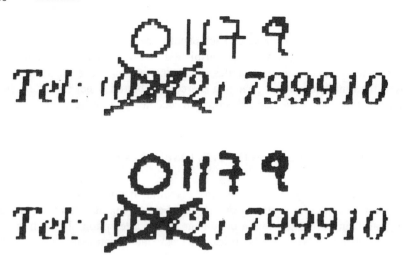

Figures 13a and 13b. Grayscaled and non-grayscaled handwriting.

look unsteady (see Figure 13a). If we had used grayscale, the writing would have looked steadier, and legibility might also have been improved (see Figure 13b). Further improvements are possible by higher resolution pen sensing or by artificially smoothing the sensed pen motion.

Like normal fax machines, DeskSlate was not designed for long-term fax storage. However, we did expect that a few documents would need to be held there temporarily, so we provided a simple visual filing system based on small pictures of documents called *thumbnails*. In the initial implementation, the thumbnail documents were arranged in neat rows, with partially overlapping pages (see Figure 14). We had planned to allow users to move the documents around more freely in later prototypes, but the simple regimented storage model seemed to work well in user trials. By using grayscale, it was possible to make individual document thumbnails recognizable despite their small size. Even though the thumbnails were less than an inch across, it was possible to see annotations that were only one pixel thick on the life-size view of the page.

## Pen Sensing and Processing Speed

In DeskSlate, the interpretation of an action as an attempt to press, draw, drag, or whatever, depends on the surface that is displayed under the pen. Any visible effect of the interpreted action must seem to result from the physical contact of the pen with that surface. The accurate alignment of user actions and display contents is particularly important during writing and drawing, when ink must seem to flow from the tip of the pen.

Pen movement is detected as a sequence of discrete points, which must be close enough together to provide an accurate representation of the pen's path.

The rate of detection and interpretation of points, and the corresponding update of the display are critical—if the device cannot convert the points into visible ink marks on the screen as quickly as they are produced, some points must be ignored, further reducing the accuracy. The conversion delay may also cause the ink marks to lag visibly behind the position of the pen. If the ink appears too slowly or does not accurately follow the path of the pen, the illusion of direct manipulation will be destroyed.

However, these problems were overshadowed by pen-calibration problems in the DeskSlate prototypes. In some areas of the screen, the actual and detected pen positions differed by over a millimeter. This sometimes prevented accurate positioning of one mark relative to another. Luckily, it did not seem to affect continuous writing severely, except when dotting an "i" or crossing a "t." In the end we had to display a tiny dot on the screen at the detected pen position to allow users to predict where marks would appear.

When people discuss documents during face-to-face meetings they often make annotations, but also point to areas of the pages without making marks. In DeskSlate, the pen position can be detected when the pen is close to the screen but not pressing it. While a document is being shared, the detected positions are used to provide feedback of the movement of the other user's pen. The feedback

Figure 14. Thumbnails of pages, arranged to convey document structures.

allows users to direct each other's attention by gesturing at the parts of the document being discussed. As for the feedback of writing, the pointer movement must not lag noticeably behind the movement of the pen. The gestures must also remain well synchronized with the verbal discussion that they accompany.

Unlike most pen-based computers, DeskSlate must process pen actions and update the display rapidly while simultaneously encoding and transmitting the actions across the telephone line and receiving, decoding, and displaying the effects of the pen actions of the remote user. Many seemingly unrelated aspects of our software were therefore designed with rapid pen-based interaction in mind.

## Liquid Crystal Display Problems

Liquid crystal displays are now the most common of all the flat panel image displays. The underlying technology is improving rapidly, but it is still relatively immature. Although most of the remaining technical problems can be overcome in isolation, there is no single display that solves all of them together. Luckily, designers can choose the displays which suffer least from the problems that affect their particular applications.

We experimented with a variety of flat panel displays before selecting one for our DeskSlate prototypes. The first decision was whether to choose a display that was illuminated from behind by a built-in light source, or one that was designed to be viewed by reflected light from the front. The differences in power consumption between the two display types were not critical to our device, but we discovered another difference that was. In DeskSlate, most of the thumbnails were likely to be predominantly white, so they would require a dark background to make them stand out. When we displayed thumbnails on back-lit displays, the thumbnails looked like small paper documents on a dark surface as we had intended. However, on reflective displays the white thumbnails looked like holes in the dark background. Adding shadows seemed to help a little but we decided that the illusion of paper was important enough to rule out reflective displays for the subsequent DeskSlate prototypes.

The surface qualities of available displays also had a significant influence on our eventual decision. If a screen is glassy and the pen nib is hard, the pen can skid over the surface, making fine control of writing difficult. Unfortunately, poor quality handwriting doesn't just look untidy—it looks childish. Even though we chose one of the best displays, some of the participants in our field trials were reluctant to send important business messages from DeskSlates because their handwriting looked unprofessional.

The surfaces of current liquid crystal displays can also reduce the illusion of direct manipulation. The liquid crystal in the display panel is trapped between sheets of glass, and for pen sensitive displays there is usually another sheet of protective glass over the top surface. A pen nib resting against the top surface of the display is a significant distance from the image in the crystals and is therefore subject to viewing parallax.

Careful interface design can help to reduce some of the technical problems with liquid crystal displays. For example, the curved edges of DeskSlate's buttons and control panels were not just an attempt to harmonize the software with the shapes of the plastic around the screen; they also arose as a solution to a display problem called *crosstalk*. On panels that suffer from crosstalk, any straight horizontal and vertical edges of displayed objects can cause ghost lines to extend over other areas of the screen (see Figure 14). Even very subtle curves were enough to avoid the problem. The crosstalk problem was yet another reason for rejecting standard graphical interfaces, with their predominantly rectangular forms.

A further technical problem is that the liquid crystals in current displays do not react instantly to changes in screen content. For example, as annotations are added to a fax page they fade from white to black and therefore seem to lag a short distance behind the pen no matter how quickly they are detected and processed. The time taken for the liquid crystals to switch state can cause even greater problems for the pointer. The pointer leaves no permanent marks, so the crystals may not switch state fully before its position changes. A fast-moving pointer will only appear to be a pale gray, making it hard to spot unless it has a bold design.

Movement poses another problem. If the position of a displayed surface changes, its previous image does not disappear instantly because of the time taken for the crystals to switch state. This affects many dynamic aspects of DeskSlate such as page scrolling and the movement of control panels. The problem is compounded by image processing delays—even a very simple shape moving over a plain background can be hard to animate smoothly with a low-power processor. The problems are worse for large or irregularly shaped objects, especially if they are moving over a complex or changing background. For fast animation, then, the gaps between successive image positions may be large and each image in the sequence may remain on the screen long enough to stand out clearly. A single object that is displayed at four different positions in quick succession can look more like four identical objects than one that is moving.

We found that the illusion of continuous motion was greater if the animations were controlled according to precise patterns of acceleration and deceleration. This was particularly important in the case of scrolling, where the movement was visible to both users. Therefore, instead of a scroll bar we provide buttons for scrolling to the top and to the bottom of a displayed page. The scroll motion automatically accelerates from rest to a maximum speed, and decelerates to rest at the end of the motion. This regular pattern of motion seems to aid mental reconstruction of the whole page, even though the page is only ever visible in fairly narrow slices. The movement starts and ends slowly. None of the changes are sudden, and the controlled acceleration ensures that the motion is predictable. Sudden changes of an interface can be disorienting (Woods, 1984), but this regular scrolling motion seems to help maintain visual continuity even for the user who did not initiate it.

## DESIGN INFLUENCES OF THE COLLABORATIVE TASK

It should come as no surprise that the DeskSlate interface has been influenced by the task for which it was designed. Nonetheless, some issues raised by the human and technological limitations of collaboration are worthy of discussion.

### Device Operation During a Telephone Conversation

Interface design standards should only be breached with good reason. The graphical interfaces of personal computers may not be perfect, but they have been standardized and are well understood by many developers and a large user population. The DeskSlate user interface does not adhere to these standards. The decision to develop an alternative interaction style was not taken lightly—it stemmed from the nature of the communication task for which DeskSlate was designed.

Unlike the majority of devices that have pen-sensitive displays, DeskSlate is not portable. The pen was not chosen to allow one hand to enter information onto a device held in the other: It was chosen for the same reason that pens are used to mark whiteboards and paper during face-to-face meetings. It is an extremely convenient communication tool.

We have already seen that existing graphical interfaces are not ideal for pen-sensitive devices, but there are two further reasons for developing an alternative interaction style. First, the device is a tool rather than a workpiece. As such, its principal function is to help its users to communicate. Their attention should therefore be focused on what is achieved via the device and not on their interaction with it. Second, operation of the device must not disrupt the process of verbal communication.

In the first case, we felt that interaction with existing graphical interfaces required a significant amount of conscious effort. When the user's only goal is to cause a change to something displayed via the interface, this is not a problem—in fact, engagement with the interface objects can be a positive benefit. But when the goals of communication are external to the device, the user's attention will be split by conscious device operation. Our aim was thus to design a tool whose interface was so simple and "ready-to-hand" that its users would hardly be aware of it (Winograd & Flores, 1986).

Our second concern was that physical or mental aspects of device operation might interfere with the process of communication. For example, if the interface had required the use of a keyboard rather than a pen, it would have conflicted with the simultaneous use of a telephone handset. Thus the tasks of device operation and telephone communication would have competed for use of the user's hands. Even with a pen, the tasks may still compete for use of the user's mental resources such as short-term memory. Research has shown that the more closely two simultaneous activities resemble each other, the more likely they are to compete for mental resources, and hence the greater the risk of interference between them (Eysenck & Keane, 1992). Interaction with a computer is often described as man-machine communication. Even when the interactions are graphical, they often constitute an ongoing dialogue, so there is a significant risk that current styles of interface will interfere with verbal communication. Our second aim was therefore to design an interface that avoided user-machine dialogues.

In response to these two concerns, we tried to design an interface with the minimum acceptable functionality, so that its capabilities and limitations would never have to be explored during a conversation. We also tried to ensure that the functions could be operated by spontaneous actions. In other words, the actions could be motivated by currently visible features of the interface, rather than by any preceding interactions.

**Supporting Unstructured Collaboration**

Telephone conversations are highly unstructured, and serve a vast range of different purposes. Although the introduction of a shared document will inevitably change the nature of telephone discussions, we did not want to impose a rigid discussion process.

We did not want to restrict the use of DeskSlate to a particular set of tasks or users either. It is not specifically for discussing financial data, or for discussing production schedules, or for discussing draft manuscripts, yet DeskSlate can support all of these tasks and more. Of course, the penalty is that it does not provide the job-specific functions that might be desired by accountants, production managers, authors, or other users.

In DeskSlate, the pen is a tool for person-to-person communication rather than data entry. As such, the ongoing pen actions are almost as important as the information in the marks left behind. The pages of a DeskSlate document are the communication medium for the ongoing pen actions as well as a permanent record of the marks. It is important for the pen actions to be synchronized with any accompanying verbal remarks. Handwriting and other pen actions need not be recognized or interpreted by the document pages; in fact, the substitution of neatly printed characters for the pen actions would strip useful timing and expression from the communication.

We have thus designed the documents in DeskSlate to be passive surfaces, like paper. They can be viewed and marked with a pen, but they do not respond in any other way to the marks that are made. The documents have surface appearance, but have no internal behavior to provide specific editing functions or to influence the course of the telephone discussion.

## Designing for Multiple Users

Annotation and erasure result from the same action on a pen, but are distinguished by a pen mode. In an interface designed for a single user, the attribution of a mode to the pen or the document would not be of major significance. However, in the case of DeskSlate the attribution is highly significant because states of the document surface are shared, whereas states of the pen are not.

If a mode is attributed to the pen, the mode will only affect the one user, but will affect that user's actions on multiple-interface objects. If a mode is attributed to a surface, it will affect both users, but only for actions on that surface. For example, a knob will rotate rather than slide no matter which user interacts with it.

Ideally, the DeskSlate pens would have had directly controllable state to determine whether they wrote thickly or thinly, or erased (Francik & Akagi, 1989). For example, one end of the pens could have been for writing and the other for erasure. Thickness could have been determined by the amount of pressure used during the manipulation. Unfortunately, we did not have time to build this functionality into the prototype. Instead, a user can press one of three buttons on a displayed button panel to select the desired mode.

Viewing modes must also be attributed to shared or nonshared state. As an example of a viewing mode, consider a document that is only half visible on the screen. If the mode was associated with a single user, each would be able to scroll the document independently. If the mode was associated with the state of the document, scrolling it would alter the portion visible to both users. In this example we designed the mode into the shared surface, so scrolling the document changes its position attributes such that both users are always viewing the same portion.

If the actions of two users can change the state of a single object, there is a risk that the changes will conflict. In the real world, conflicting actions on a surface are resolved immediately by physical forces such that a single consistent effect is produced. In DeskSlate, each user must be able to manipulate a surface and see the direct effect instantly. By necessity, for the effects to be instant they must also be independent. Therefore, if both users can see the same document, they can act on it simultaneously in ways that produce inconsistent direct effects.

Conflicts must be detected and resolved such that the final appearance of the shared surface is the same for each user. The resultant state may not be as anticipated by the users, but should seem to be a natural consequence of the conflict. For example, if one DeskSlate user scrolls a document while the other is sketching on it, the resultant sketch will be split between the original and final scroll positions, with a straight line inbetween. The same result might have occurred if similar actions had been performed on a real paper document. In most cases, it is easier for users to recover from an occasional conflict than to cope with an interface which has been designed to prevent conflicts happening. However, if users want to prevent conflicts, they can devise their own strategies to do so. For example, we have seen users draw lines to divide the shared surfaces into smaller regions for individual use.

### Data Transmission Delays and DeskSlate's Shared Area

Many shared workspaces have been built for collaboration over data networks. DeskSlate was built to provide similar capabilities during ordinary telephone calls. The technical limitations of telephone-based data transmission had to be considered when designing the DeskSlate interface.

Current telephone networks have been designed primarily for voice calls, but they can also be used by fax machines and computer modems to transmit electronic data. Transmission speeds tend to be much lower than those seen over networks designed specifically for data, but in compensation the telephone networks are more widely accessible from homes and businesses.

The treatment of shared documents in the DeskSlate interface has changed with successive prototypes. Initially, the same actions allowed a document to be viewed, whether or not a telephone call was in progress. Between calls, the document would become visible on the local machine for private editing. During calls, the selected document would become visible on both machines for shared editing. Although the actions may have been the same in both cases, in the shared mode of operation there could be a long delay before the document became visible to both conversants. The delay was suffered once for each page shared during a call, while a copy of the page was transmitted over the telephone line. Thereafter, the conversants could switch rapidly from viewing one shared page to another.

Users could learn that the first attempt to view a page during a call would involve an extra long delay, but there was nothing in the original interface to help explain or predict the effect. We planned to differentiate the thumbnails of pages which had already suffered the transmission delay (see Figure 15), but the differentiation would not have made the viewing delays any easier to understand. Instead, a shared document area was introduced into later prototypes to convey the relationship between sharing and the delays.

The shared area was a distinguished area of thumbnails (see Figure 16). Each person could get a clear view of the documents available for discussion, whether still in their private storage or already in the shared area. During a call, both participants would be able to select documents to share, and would see the thumbnails appear in the shared area at the end of the transmission period. At the end of the call, the shared area would automatically empty, and copies of the shared documents would appear in each person's private storage area.

Previously, viewing a document during a phone call had made the document visible to the other person as well. In the modified scheme, users had to move stored documents into the shared area to make them available to the other person. The transmission delay was associated with the entry of documents into the shared area. Therefore, the action of viewing a document was free from delay at all times.

The new model of operation did not provide a more truthful explanation of the delay, but it was consistent with having a single delay for each page shared during a telephone call. In reality, a document is sharable if a copy exists on the other DeskSlate. Then the document state can be displayed to both users, and only the incremental changes need to be communicated over the phone line to update the displays. As a concept, sharing is simple, but the reality of its implementation is too complex for the user interface. Shared areas provide a visualizable alternative to both the abstract concept and the complex reality.

The results of tests of the shared area were promising but not entirely successful. The main problem was that the original shared area appeared to be on the same local surface as the thumbnails stored in the DeskSlate. Consequently it did not convey the distinction between shared and nonshared thumbnails very well, and did not help to explain the time taken to transfer a document to the shared area. An idea for an improved shared area came while considering a seemingly unrelated enhancement.

Research has shown that being able to see the movement of other people's cursors can enhance the use of a shared workspace (Tang, 1991). If someone knows that their cursor can be seen, they will tend to use it as a gesturing tool. For example, they might move it in a circular motion around a displayed item that they are discussing. A mechanism had been provided in DeskSlate for gesturing, but only while a shared page was being viewed. Because shared and nonshared areas of thumbnails were adjacent, how could we support gesturing over one and not the other? A simple solution was to display the shared area through a hole in the local surface (see Figure 17). The other person's cursor would then come into view when it was over the shared area, but would pass under the local thumbnails at other times. As it turned out, this also addressed the problem of showing the shared area without it seeming to be part of the local device.

## A Visual Interlock For Controlling Telephone Line Status

To make a phone call using a DeskSlate prototype, users can dial in the normal way with the attached telephone. When the call is answered they can begin a

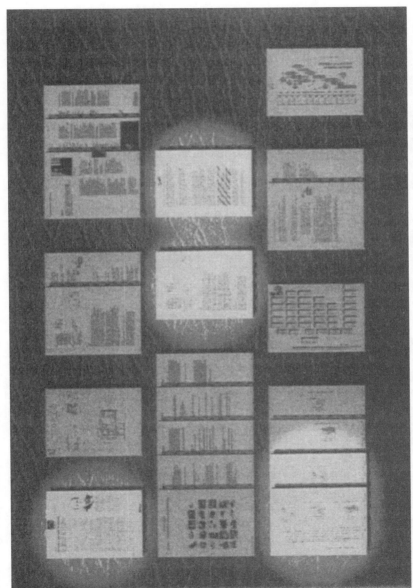

Figure 15. Shared pages indicated by pools of light.

Figure 16. Shared pages in a separate shared area.

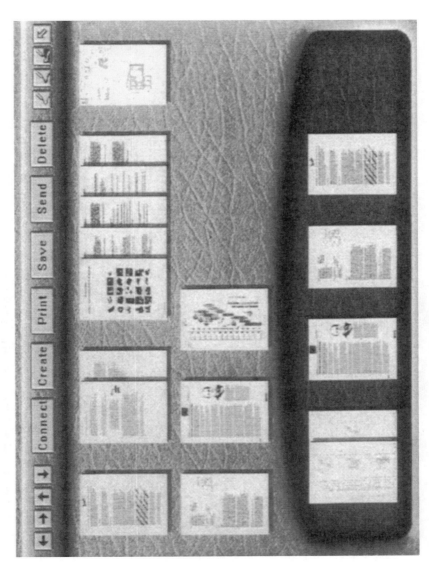

Figure 17. Initial design for a recessed shared area.

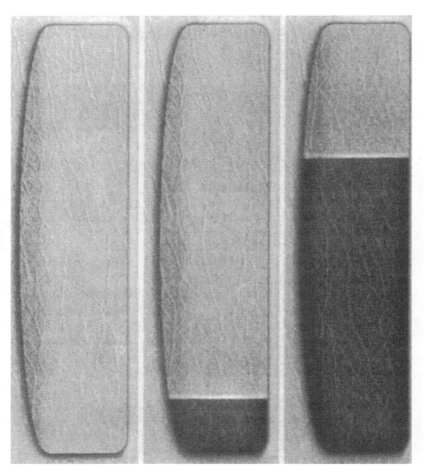

Figure 18. A shutter slides back to reveal the shared area.

normal telephone conversation. If the other person also has a DeskSlate proto-type, they can enable a mode in which sharing is possible by pressing a button labelled "connect." At first, it was too easy for users to forget to press this button before trying to share a document. The interface was thus modified to overcome the problem.

Actually, this is very similar to a problem that existed with early tele-phones. Lifting the telephone earpiece off a hook in order to answer a call did not signal to the telephone exchange that the connection should be established. A separate action was required, yet that action achieved an effect which was never fully understood by telephone users, and was easy to forget. Later tele-phones disguised the action of establishing the connection by mechanically linking it to the action of lifting the handset. The hook became a switch that was automatically triggered by removing the handset. The new hook avoided users having to be aware of the connection process, and also ensured that they could not answer the call without establishing the connection first. Due to this innova-tion, the states of the telephone line that signal the intention to make or accept a connection are now known as *on-hook* and *off-hook*.

Inspired by the early telephone designers who linked the line connection status to the more obvious status of a handset being on or off a hook, we decid-ed to link the shared status of our DeskSlates to a conspicuous visible state of the interface. No appropriate visible state existed, so we introduced a new one. We added a shutter to hide the shared area until sharing was enabled. Pressing the "connect" button then enabled the shared mode as before, but also caused the shutter to slide back to reveal the shared area (see Figure 18). Therefore, the removal of the shutter provided conspicuous and relevant feedback of the initia-tion of the shared mode, but more significantly for users, the presence of the shutter provided a motivation for initiating the mode in the first place.

We have now positioned the "connect" button on the shutter itself to pro-vide a more conspicuous link between their behaviors (see Figure 19). The new button behaves like a release mechanism to open the shutter. A user presses it, the shutter slides off the top of the screen, and the shared area becomes avail-able for use (see Figure 20). At the end of the call, or if the sharing mode fails due to a poor quality phone line, the shutter drops back into its original position. Thus the button to initiate the mode is conveniently hidden while its use would be inappropriate, and the shutter and shared area give graphic feedback of the initiation, availability and eventual cessation of the sharing mode. One of the rules of interface design is that all interface behavior should have a plausible source, even if the behavior results from a user or system "error" (Lewis & Norman, 1986). So in DeskSlate, the behaviors are related to specific surfaces, or to the imaginary internal workings behind the surfaces. The shutter provides a good example of this. Having repositioned the button, everything to express and control the shared status of the telephone call is now clearly related to the shutter. Even the error conditions are expressed through its visible behavior.

*Interlocks* are mechanical links between device states and control surfaces.

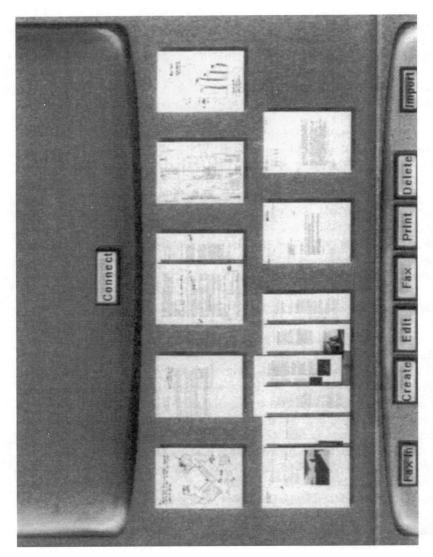

Figure 19. Connect button on the shutter of the current prototype.

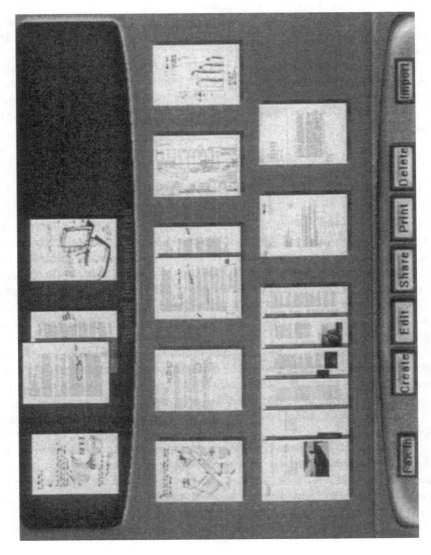

Figure 20. The shared area of the current prototype.

They either block an inappropriate user action, or automatically switch the device into a state where the action is appropriate. Consider a telephone with its keypad hidden under the handset. The physical obstruction of the keypad is an example of the first type of interlock. It prevents dialling while the handset is on the hook. The hook itself is an example of the second type of interlock. The action of lifting the handset off the hook initiates a state in which it is possible to dial. The DeskSlate shutter behaves in a similar manner to the telephone. When the shutter is closed over the shared area, it discourages premature attempts to share documents, and when it is opened, it automatically triggers a state in which the documents can be shared. However, notice that the shutter does not block the attempts to share a document; rather, it seems to block the visualization of their success. The technique we have used could therefore be more appropriately termed a *visual interlock*.

This Visual Interlock has proved to be a useful user interface mechanism. In user trials, it has deterred actions that would have been inappropriate for the current status of the telephone line, and also allowed users to observe and control the line status without any understanding of the underlying communications technology.

## SUMMARY AND DISCUSSION

This has been a case study of the role of constraints in the activity of design. It has focused particularly on the constraints imposed by pen-based interaction and document sharing, which are emerging as two major new areas of research. We have described an iterative design process, with an initial focus on the exploration of human and technological barriers via disposable design solutions. The focus gradually shifted to smaller and smaller design details, but the process remained one of exploration, rather than progressive refinement of a single solution.

While describing some of the specific influences of the technologies of pen-based interaction, we have picked out examples of interface elements that were designed to overcome unexpected technology barriers. Without deliberate exploration, we could not have anticipated the extent to which the technologies of implementation would affect both the broad concepts and the fine details of the design.

Of course, design issues are rarely independent, so all decisions involve a degree of compromise. A solution to one problem may introduce another new problem elsewhere in the design; however, we also found that by exploring a range of solutions to each problem we would sometimes stumble on ways to overcome other seemingly unrelated problems.

In the third section, we described some of the influences of pen-based interaction and our responses to them. In order to provide many possible effects from the small number of distinct actions possible with a pen, we divided the user interface into distinct surfaces which visibly conveyed the manipulations that they afforded. Pen-operated controls were then described as an illustration of interface elements constructed from manipulable surfaces, and as an example of the adaptation of existing interface elements to suit pens. The evolving designs of buttons illustrated the careful design necessary to convey the surface affordances.

In the fourth section, we discussed the design influences of the display technology. Legibility of documents is critical to a device such as DeskSlate, yet we must work within the constraints of current display resolutions. We described the use of a grayscale display to compensate for the low resolution. For remote collaboration, the dynamic aspects are almost as important as the static presentation of documents. In particular, smooth handwriting requires very close integration of sensing and display technologies with rapid feedback from one to the other. Other display issues were overcome by appropriate user interface design. For example, the problem of display crosstalk was overcome by the use of curved surfaces in the interface.

Some of the design influences of the collaborative task were discussed in the fifth section. We found that the fundamental style of the user interaction had to be geared for use during a telephone call, and that the functions of the device had to be general enough to support vast range of different collaborative activities. Interface concepts such as the Shared Area and Shutter were introduced to express constraints imposed by the telephone network. The shutter is an example of a visual interlock. It provides a simple means to ensure that users perform actions in a specific order.

### Models of Human–Computer Interaction

In Norman's (1988) model of the human–computer interaction process, the user evaluates the state of the interface, conceives a goal, and executes an action on the interface to achieve the goal. If the interface provides feedback, the user evaluates it to begin the cycle again. The action and feedback are inherently linked by the interface state (Draper, 1986). When representative display of information is combined with simple and representative control (Buxton, 1986), the result is an enhanced feeling of direct engagement with the objects in the interface (Laurel, 1986).

Norman's (1988) model provides an appealing framework for guiding interface design. DeskSlate requires some human-computer interaction of the sort described by the model, but is primarily for human-human interaction. However, the two need not be very different. Users can communicate with each other by the effects of their actions on shared interface state (Sorgaard, 1988). This has become an area of much recent research, especially when combined

Figure 21. The HP Omnishare interactive fax product.

with verbal communication (Bly, 1988).

For many actions such as writing and drawing, immediate feedback is essential. Without it, hand-eye coordination is impaired. The model works well for the interactions that have immediate visible effects on the manipulated point of the interface; however, there are also many effects that involve processing delays or require responses from other users. Those effects are not so well accommodated by the model because they do not result from direct manipulations alone. The effects also depend on the system behaviors behind the interface, or on external behaviors of other users. Such behaviors may break the interaction cycle by delaying effects or displacing them to another point on the interface.

In DeskSlate, we have ensured that all actions have an immediate effect at the point of contact with the pen, even when the users' primary goals are to trigger a delayed or displaced indirect effect. For example, when users press a button on the surface of an object to trigger an internal behavior, they will always see an immediate change to the button whatever the internal behavior.

## Feedforward

Since studies of cybernetics began in the 1940s, *feedback* has been recognized as a vital process in the regulation of human motion. Consequently, feedback mechanisms are now considered essential in user interface designs. In 1967, Bernstein suggested *feedforward* as an essential complement to feedback. For example, the motion necessary to catch a ball cannot be explained by feedback alone; it must also be based on a predicted path of the ball ahead of its observed position. This is an example of feedforward by extrapolation of observations. To explain voluntary action, as opposed to reaction, Bernstein proposed an additional form of feedforward based instead on interpolation between an observed state and a desired future state.

Visible affordances of surfaces can guide users to apply appropriate surface manipulations (Gibson, 1977), but where users have goals to achieve indirect effects, they must have effective mental models of the behaviors behind the surfaces (Gentner & Stevens, 1983). Mental models are not static descriptions of objects. They are active mental simulations that can be subjected to imaginary manipulations to foresee the direct and indirect effects. In that sense they are consistent with Bernstein's (1967) notion of interpolative feedforward.

Studies of feedforward provide valuable insights into psychological aspects of interaction. For example, they help to explain the mechanisms that underlie situated actions (Suchman, 1987). In the DeskSlate interface, we have tried to provide appropriate feedback mechanisms to convey the effects of users' actions, but more significantly, we have tried to design the surfaces to provide sufficient feedforward mechanisms to motivate the actions in the first place.

## What Has Come of DeskSlate?

The appearance and functionality of the DeskSlate interface could be improved significantly, but it was only intended as a prototype and will probably not be developed any further. Yet even as it is, DeskSlate has been remarkably successful. We believe that we have achieved the goal of designing an interface for unobtrusive operation, and we have learned a lot in the process.

What happened to the concept of an interactive fax? As a parallel activity during much of the development of DeskSlate, we liaised with a product development team in another part of the company to support plans to develop an interactive fax device. As a result, many of the lessons from our research, and some of the design elements, have found their way into the HP OmniShare, a real interactive fax product lauched recently in the United States (see Figure 21).

## REFERENCES

Benbasat, I., & Todd, P. (1993). An experimental investigation of interface design alternatives: Icon vs. text and direct manipulation vs. menus. *International Journal of Man-Machine Studies, 38*, 369–402.

Bernstein, N. A. (1967). *The coordination and regulation of movements.* Oxford: Pergamon Press.

Bly, S. (1988). A use of drawing surfaces in different collaborative settings. In *Proceedings of ACM CSCW `88,* pp. 250–256.

Buxton, W. (1986). There's more to interaction than meets the eye: Some issues in manual input. In D. A. Norman & S. W. Draper (Eds.), *User centered system design: New perspectives on human-computer interaction* (pp. 319–337). Hillsdale, NJ: Erlbaum.

Cushman, W. H., & Miller, R. L. (1988). Resolution and gray-scale requirements for the display of legible alphanumeric characters. In J. Morreale (Ed.), *1988 SID International Symposium. Digest of Technical Papers* (pp. 432–434). Playa del Rey, CA: SID.

Darke, J. (1979). The primary generator and the design process. *Design Studies, 1*(1), 36–44.

Draper, S. W. (1986). Display managers as the basis for user–machine communication. In D. A. Norman & S. W. Draper (Eds.), *User centered system design: New perspectives on human-computer interaction* (pp. 339–352). Hillsdale, NJ: Erlbaum.

Eysenck, M. W., & Keane, M. T. (1992). *Cognitive psychology: A student's handbook.* Hillsdale, NJ: Erlbaum.

Francik, E., & Akagi, K. (1989). Designing a computer pencil and tablet for handwriting. In *Proceedings of the Human Factors Society 33rd Annual Meeting* (pp. 445–449). Santa Monica, CA: Human Factors Society.

Gentner, D., & Stevens, A. L. (Eds.). (1983). *Mental models.* Hillsdale, NJ: Erlbaum.

Gibson, J. J. (1977). The theory of affordances. In R. E. Shaw & J. R. Bransford (Eds.), *Perceiving, acting, and knowing* (pp. 67–82). Hillsdale, NJ: Erlbaum.

Kobara, S. (1991). *Visual design with OSF/Motif.* Reading, MA: Addison-Wesley.

Lambert, S. (1993). *Form follows function?* London: Victoria & Albert Museum.

Laurel, B. K. (1986). Interface as mimesis. In D. A. Norman & S. W. Draper (Eds.), *User centered system design: New perspectives on human-computer interaction* (pp.67–85). Hillsdale, NJ: Erlbaum.

Lewis, C., & Norman, D. A. (1986). Designing for error. In D. A. Norman & S. W. Draper (Eds.), *User centered system design: New perspectives on human-comput-*

*er interaction* (pp. 411–432). Hillsdale, NJ: Erlbaum.

MacKenzie, I. S., Sellen, A., & Buxton, W. (1991). A comparison of input devices in elemental pointing and dragging tasks. In *Conference proceedings CHI'91* (pp. 161–166). New York: ACM Press.

Nielsen, J. (1993). *Usability engineering.* New York: Academic Press.

Norman, D. A. (1988). *The psychology of everyday things.* New York: Basic Books.

O'Conaill, B., Geelhoed, E., & Toft, P. (1994). DeskSlate: A shared workspace for telephone partners. In C. Plaisant (Ed.), *Conference Companion, CHI'94* (pp. 303–304). New York: ACM Press.

Owen, D. (1986). Naive theories of computation. In D. A. Norman & S. W. Draper (Eds.), *User centered system design: New perspectives on human-computer interaction* (pp. 187–200). Hillsdale, NJ: Erlbaum.

Sorgaard, P. (1988). Object oriented programming and computerised shared material. In S. Gjessing & K. Nygaard (Eds.), *ECOOP `88 Conference Proceedings* (pp. 319–334). Berlin: Springer-Verlag.

Suchman, L. A. (1987). *Plans and situated actions: The problem of human-machine communication.* Cambridge: Cambridge University Press.

Tang, J. C. (1991). Findings from observational studies of collaborative work. *International Journal of Man–Machine Studies, 34,* 143–160.

Whittaker, S., Geelhoed, E., & Robinson, E. (1993). Shared workspaces: How do they work and when are they useful? *International Journal of Man–Machine Studies, 39,* 813–842.

Winograd, T., & Flores, F. (1986). *Understanding computers and cognition.* Norwood, NJ: Ablex.

Woods, D. D. (1984). Visual momentum: a concept to improve the cognitive coupling of person and computer. *International Journal of Man–Machine Studies, 21,* 229–244.

# Chapter 6
# User Knowledge and User-Centered Interface Design in Microsoft Applications

Ken Dye
Chris Graham

*Microsoft Corporation*

Before 1989, most commercial software development entailed applying computing power to respond to customer requests and competitive challenges. The so-called "feature wars" made sense, given the state of the software industry. Software technology had not yet solved many user problems. Users wanted products with features that allowed them to accomplish an ever-widening variety of tasks, and personal computer hardware was increasing in power. As a result, software developers were presented with many opportunities for design that did not involve carefully studying users and their work. The burden on developers was not finding new design opportunities but creating technology to solve existing problems.

At that point in the evolution of personal-computer software, Microsoft devoted fewer resources to user-interface issues than we now do for two reasons:  First, designers, constrained by short product cycles, worked on issues deemed most important to customers—the backlog of existing design ideas was large, and our first priority was to address that backlog; second, rapid and straightforward advances in personal computers meant more opportunities to take advantage of the power of Windows and graphical-user-interface applications than we had the resources to pursue.

Beginning with Microsoft Excel 3.0, we began to devote more design and development resources to usability issues and user-interface design. We understood that improving our products' user interfaces required a better understanding of users and the environments in which they work.  In 1988 we began to rely more heavily on empirical methods for discovering relevant design constraints.  We spent more time working with users in their workplaces and met with customers in MIS departments. We developed and instrumented versions of our products and distributed them to users in the field to capture highly detailed information about users' interactions with our applications. We also did much more work in our usability lab to evaluate prototypes. Over the course of the last few years, such investigation has become a routine part of the design and development process.

Our goal was to design products tailored to the real activities done by people as they do their work. Good design is a combination of power and the user's ability to apply that power. It is difficult to separate what users can accomplish with a product from how easily they can accomplish it—to separate utility from usability. We did not see product design and user-interface design as separate issues, and the challenge we faced was to make them part of the same process.

## USER KNOWLEDGE AND USER-INTERFACE DESIGN

User-interface design at Microsoft is driven by the need to satisfy many constraints. We must pay attention to the current state of hardware, to release schedules, to the trade press, to customer requests, to competitors' products, and so on. But by far the most important constraints are generated by our users—by the work they do, the goals they bring to their work, and their knowledge of the products they use.

Designing software to help users solve the problems they confront outweighs other design constraints. We pay careful attention to performance issues such as visual acuity and manual dexterity and also attend to cultural issues such as which images are most recognizable and which vocabulary is most appropriate. However, given the degree to which using a computer is a cognitive activity, we believe that the most important design issues to take into account have to do with the user's knowledge and how they employ it to accomplish their goals. Because of this emphasis we began to look at ways we could supplement what users know by making our products more intelligent.

Intelligent user interfaces are typically described as interfaces that take information from the user's behavior or expert domain knowledge (Hefley & Murray, 1992), and either change the way the software works (adaptive user interfaces) or perform some specialized task for the user (intelligent agents). The idea of intelligent user interfaces is very powerful and promising, but intelligent interfaces have yet to deliver the potential advantages. Our experience is that interfaces, by taking advantage of what we know or can learn about user behavior, can be intelligent without complex algorithms or incorporating rule-based knowledge.

Designing software requires knowing what users want to accomplish, in what environments, and based on what experience and knowledge. The framework expressed by Card, Moran, and Newell (1983), *Knowledge + Inputs + Goals = Behavior*, is a useful way to describe these relationships. Users' goals arise from the environment in which they work and live. Their knowledge comes from their experience in those environments, particularly their experience with computer software. An interface is set of predictions about user knowledge and behavior. Users know x and have these goals. If we give them these inputs, they will behave as we expect.

This view fits with our own experience with the software design process. Design at Microsoft involves a lot of discussion, with input from many sources. When we examined the nature of design discussions we discovered that almost all can be characterized by a few questions:

- What does the user want?
- What does the user expect?
- What will the user do?
- How does this design work?

These questions roughly correspond to, or at least encapsulate the Card, Moran, and Newell formula. Asking "What does the user want" is the same as asking "What are the users' goals." "What does the user expect?" corresponds to the questions "What do users know, what kind of inferences do they make about how the interface works based on this knowledge?" "What will the user do?" is a question about user behavior. "How does this design work?" is a question about inputs. The designer's job is to design successful inputs, in the form of a successful user interface, that leads to productive problem solving. The designer is saying "I designed this interface this way because I predict users will be successful with it. How good are my predictions?" Therefore, both our theoretical understanding of the design process and our practical experience tell us that our focus during the design process should be on user goals and user knowledge. By focusing on users' goals and knowledge we have found many opportunities for designing "intelligent" interfaces.

## What Users Know and User-Interface Evolution

The problem of user knowledge is particularly difficult when designing packaged software. Every product we ship must take into account the millions of users who have invested time in adapting that product to their environment and who have developed strategies for using the product to accomplish work. Furthermore, users have developed a rich intuitive knowledge of the natural physics of the world, and how to use the artifacts and tools of society. The challenge in designing the user interface for products such as Microsoft Excel and Microsoft Word is to take this knowledge into account and to extend it. Our goal is to make products that both take advantage of users' accumulated knowledge and are more intelligent about users' knowledge and users' goals. By more intelligent about user behavior we mean products that anticipate user behavior:

1. Based on data collected from users
2. Based on data the user has entered
3. Based on goals the user has selected.

Thus, if we can successful anticipate user behavior, we can either do useful work for users or provide users with the information they need to do useful work.

## HCI Research and Commercial Software Design

As we began to confront more difficult user-interface design problems, we began to look to human–computer interaction research as a guide to designing intelligent interfaces. We found, however, that HCI research did not always address our specific needs. As software designers, we need answers to specific questions quickly, often within a few weeks. Researchers often look for more reliable and generally applicable answers to larger questions—answers that can become part of a broader understanding of human–computer interaction.  In some cases, researchers attempt to develop new technologies that might be useful for long-term design and development goals but that are not applicable to the next release of a product. As a result, it is difficult for us to point to specific research ideas that have led to specific design changes.

Byrne, Wood, Sukaviriya, Foley, and Kieras (1994) captured the problem in a paper presented at CHI '94 in a way that reflects our experience:

> Often, it appears as if the field of HCI is divided into two camps, the technologists and the psychologists. Despite the efforts at interdisciplinary partnership, most real interfaces are fundamentally technology-driven. One reason for this, perhaps, is that simply developing working code is difficult enough without having to worry about the most recent findings in the psychology of human–computer interaction. (p. 232)

Despite our focus on the practical problem of shipping products, we have found HCI research to be useful and informative and we are finding new and better ways to incorporate the results of HCI research into our design process.

We have found HCI research to be helpful in three areas. First, it provides a general framework for thinking about design problems. Research in areas ranging from human problem solving to situated action (Suchman, 1987) have influenced how we think about software design. Second, research in software use has given us many fruitful ideas. Work by Nardi and Zarmer (1993) on visual formalisms and spreadsheet use is an example. Work by Johnson and Schneiderman (1991) on hierarchical information structures is another. Finally, research in usability evaluation by Nielsen (1993), Wright and Monk (1991), and Whiteside, Bennett, and Holtzblatt (1988) has influenced the way we use empirical data to support the design process.

We have not, however, found HCI research to be directly applicable to our specific, everyday design problems. We believe that the gap between software design and HCI research is due in part to the aforementioned difference in orientation, in part to historical differences, and in part to the fact that many of the problems associated with software development have to do with specific products. We have begun to consider ways to close that gap by becoming more involved in the HCI community, providing internships for students, and increasing our own HCI research efforts. However, to date it is often difficult for us to point to many specific research findings that have contributed directly to a specific design.

## ANTICIPATING THE USER'S GOALS AND ACTIONS BASED ON DATA COLLECTED FROM USERS

When we first began to look for ways to improve the user interface for Microsoft Excel 3.0, we were struck by the results of a study by Napier, Batsell, Lane, and Guadagno (1990) at Rice University regarding the frequency of command usage in spreadsheets. This study brought home a simple but important lesson—that users do some activities over and over again. We thus began to look at simplifying the activities users frequently perform.

We examined common user activities with an eye to simplifying the activities themselves. Our goal was to reduce what the user needed to know as well to minimize the number of actions the user had to perform to accomplish common activities. By observing the behavior of a large number of users, we found we could fairly accurately predict user goals within the context of these common activities. Our research in this area led to the development of heuristics that combine a number of steps into a single action, producing reliable results. AutoSum is a simple example of such a heuristic.

### Anticipating User Goals—AutoSum in Microsoft Excel 4.0

In our work on Microsoft Excel 3.0, we discovered that formulas consisting simply of the sum of a range of cells in a row or column make up a high proportion of all formulas entered in a spreadsheet. This suggested to us the idea of including the AutoSum button on the toolbar, which we hoped would make summing easily discoverable and recognizable, and would reduce the number of steps users had to perform to sum a column or row of numbers.

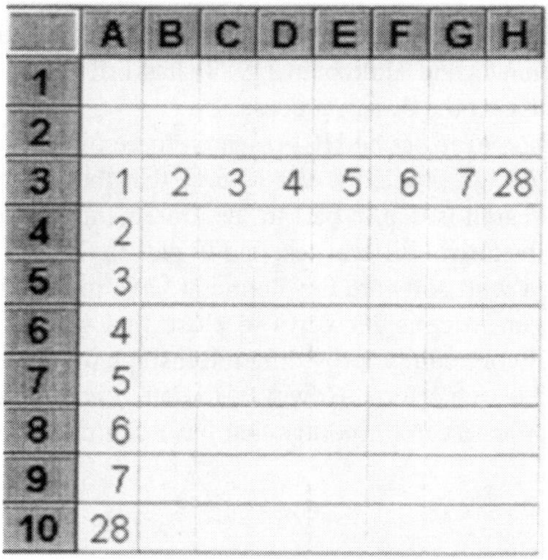

Figure 1. Results of an AutoSum.

Our first test of AutoSum involved the simplest case—summing a column or row of contiguous numbers in the cell immediate adjacent to the column or row (see Figure 1). Usability tests showed us that the AutoSum button was very successful for the simplest case, with two caveats. First, users had more difficulty interpreting the button image than we had hoped, although they easily recalled its purpose. More recent research has convinced us that including such a button was the right decision—that for long-term efficiency of use, how easily users recognize and recall an item are more important than the ease with which they initially interpret an item.

Second, we learned that addressing the simplest case of summing a row or column was insufficient. For AutoSum to be useful, it had to cover a large percentage of users' summing tasks. Otherwise users would have to learn two or more methods. To accommodate cases in which the numbers to be summed were not contiguous or the cell containing the sum was not adjacent, we included "semiselect" to allow users to modify the selection. For example, in Figure 2, AutoSum cannot decide what results to produce. By allowing the user to press the AutoSum button and then modify the selection, the user can control the result, as seen in Figure 3.

By combining a number of steps into a single operation, we had, for some cases, created what Norman called a mode error (Norman, 1988). The AutoSum button combines two functions into a single action. It selects the object to be acted on—a range—and it supplies the action—a formula. However, when the range is not contiguous AutoSum can produce an error. Because of the large number of cases in which AutoSum is useful, it is worth introducing the possibility of an error; by introducing semiselect we made that error reversible.

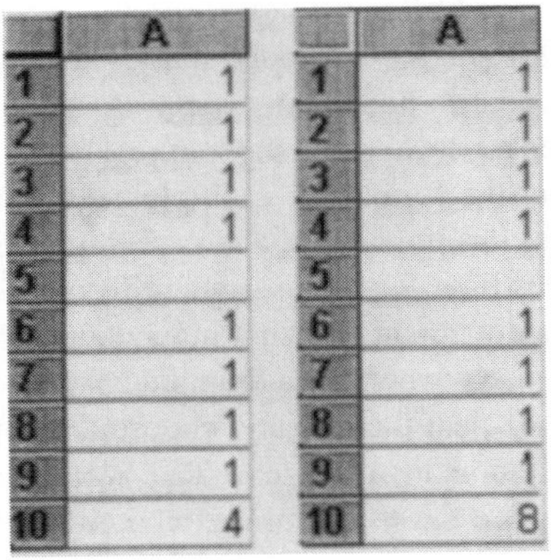

Figure 2. Which result did the user intend?

**Anticipating User Actions in Word 6.0**

One of our goals for Word 6.0 was to do more than make features available to the user. We also wanted to make Word conform to how users work. Our initial intention was to find out how users compose documents and then use that information to help them by doing some of the work. To discover how users compose documents we analyzed many, many documents and watched users create documents. We discovered that the problem was not so simple. The documents users produce do not conform to standard types, but adapt to the users' purposes and the constraints of the environment in which they work, and the process of creating a document involves many iterations. Users do not so much design documents as evolve them. As a result we began to look for ways that we could do some of the low-level work for users and allow them to focus on producing a document.

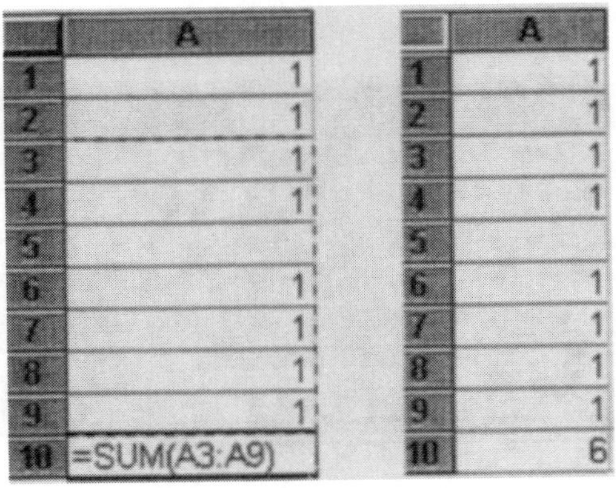

Figure 3. By modifying the selection the user can produce any result.

*AutoFormat in Word 6.0.* We decided that even if we could not determine in advance what a "standard" document type was, we could examine the text and identify text elements. Because of the nature of text it is possible to make reliable inferences about the format for each element and apply styles. Almost every document includes headings, lists, and paragraphs. Styles are a useful formatting device that saves time for the user when formatting these elements and ensures consistency in a document. Formatting a document, however, still takes time for the user, particularly when the document changes frequently as the user creates it. The user must select the text and choose the appropriate style from a dropdown list for each element of the text. The Word Development team reasoned that Word could save the user time by recognizing the text elements and automatically applying styles.

In Word 6.0 the user can format the entire document or a selected portion of the text by choosing *Format-AutoFormat* from the Menu bar or by clicking the AutoFormat Tool on the Toolbar. Word scans the text and identifies the text elements that are formatted in the Normal or Body Text styles and applies the styles from the template currently attached to the document. If a style has already been applied to a text element, Word will not apply a new style. Word assumes that the user intended to apply the style and wants to keep it. The user can reformat the entire document by clearing the Previously Applied Styles check box on the AutoFormat tab.

We were concerned about how users would react to AutoFormat, and we wanted to give users control over the formatting. Thus AutoFormat allows users to review the changes to their document. When Word has finished formatting the document, it gives the user the opportunity to accept, reject, or review the changes (see Figure 4). If the user chooses to review the changes, Word highlights each change with color and walks the user through the changes (see Figure 5).

While designing AutoFormat we were very timid about the kinds and

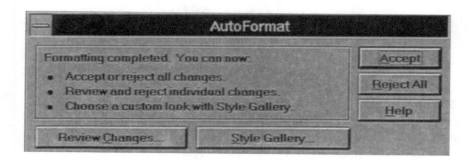

Figure 4. The user can accept, reject, or review the changes.

Figure 5. AutoFormat allows users to review the revisions.

extent of formatting we did automatically. We are very reluctant to design features that the user does not control. As a result, AutoFormat recognizes and formats fewer text elements than the algorithm will allow. For instance, in a standard document AutoFormat recognizes headings, lists, and paragraphs. When we tested AutoFormat in the Usability Lab we regretted our caution—users had little trouble using AutoFormat and were very adept at reviewing and accepting or rejecting formats.

*AutoCorrect in Word 6.0.* Not only do users frequently perform certain tasks, they also frequently make the same errors. In many cases it is not worth the time required to fix those errors. Many documents that users create don't need to be perfect—they need to be timely. Memos, notes, and e-mail messages are good examples. The user does not want to invest the time and effort into correcting errors. In other cases correcting errors can be costly. Some users tend to pause to correct every typographical error, rather than waiting to correct them later using a spell checker. This tends to slow down the composition process by forcing users to concentrate on correcting errors rather than planning and composing their document. We felt that by discovering and correcting errors for the user we could help solve both problems. We set out to find out what errors we could automatically correct.

By talking with users we discovered a number of things that really annoyed people. For example, when users select text beyond the paragraph mark, Word deletes the text, but not the paragraph mark. Another example is the "sticky shift key." When users are typing quickly, they often put two capital letters at the beginning of a word, as in "OMaha." We also discovered that these things often occur when users are producing documents quickly without a lot of concern for correctness and format.

By compiling problems users were aware of, and by examining a large number of documents, we developed a long list of items we wanted to automatically correct and culled the list for the most likely candidates. We eliminated things from the list for which we didn't think we could be right at least 95% of the time. For instance, we thought about always converting numbers like 3/4 to fractions,     . We also considered capitalizing both days of the week and months, but we rejected months because "may" and "march" could have other meanings.

We created a prototype that represented an approximation of the what we wanted it to do and thoroughly tested it with users. We learned that our prototype didn't catch enough errors and so we added a number of things to our list. Some of the items we added required that we make our algorithm much more sophisticated. For example, the initial prototype couldn't correct errors that involved improperly using symbols that were also Word delimiters. Word did not correct "don;t" to "don't" because it interpreted the semicolon as a delimiter.

The first approximation also did not handle cases in which users wanted to type something that Word interpreted as an error. For example, a user making a list of compact disks would discover that Word would automatically correct "CDs" to "Cds." Users could turn off that specific AutoCorrect rule, but that meant that Word wouldn't correct the mistakes the user wanted corrected. Or, the user could choose *Undo* immediately, but that required users to stop what they were doing and turn their attention to "fixing the correction." Our solution was to do the correction only after the user pressed the spacebar. To prevent Word from making the correction, users could use the arrow key rather than the spacebar.

One of the advantages of AutoCorrect is that it can improve the performance of spell checking. When users run the speller they can focus on correcting true spelling mistakes rather than fixing typos. We considered adding all corrections that the user did while spell checking to AutoCorrect, but we decided not to for several reasons. First of all, errors discovered during spell checking are often words not frequently used, and AutoCorrect works best for words frequently used and mistyped, or for shorthand for frequently used but hard-to-type or difficult to spell words (such as "equivalent" or "tomorrow"). We also didn't want to crowd the AutoCorrect list with too many misspellings, and we wanted to avoid the risk of conflicting replacements.

In addition, we considered the problem of different languages. AutoCorrect does not look at the language property of the text as the speller does. We didn't want to get French spelling corrections automatically entered in an otherwise English AutoCorrect. Second, users think of typos and misspellings as two different classes of problems. Typos are slips (Norman, 1988)—words users know how to spell, but their fingers trip. Spelling mistakes are those words that users don't know how to spell. Finally, we found that users often don't use the speller at all, especially when they're writing technical documentation where many words would be flagged. This makes spelling slow and cumbersome, and people often don't do it. We still wanted to provide a method to fix those annoying typos without forcing the user to run the speller.

We named the feature *AutoCorrect* because we wanted to create a consistent picture of the three related features—AutoFormat, AutoCorrect, and AutoText (AutoText was formerly called the Glossary. AutoText allows users to store and use chunks of text. Since the release of Word 6.0, we have discovered that people use AutoText much more often than they used Glossary because it is much easier to discover, and because the name helps users predict what the feature will do.)

**Tool Tips—Making Information Available to Users**

Graphical user interfaces, by making functionality visible, allow users to rely on cued recall rather than simple recall. Instead of having to learn and recall all of the available commands and their locations, users of graphical user interfaces can rely on visual elements as cues. As Norman (1988) pointed out, if the visual elements are well designed, there is little for the user to remember. This advantage is lost, however, when commands are not immediately visible. When menu commands are not visible, for example, users have to search the menu structure. Thus, designers working with graphical user interfaces attempt to make as many functions available in the interface as possible, given limited screen space.

Icons are examples of visual cues that can be available to the user. However, for icons to be useful as visual cues they must be easy to interpret, and they must be easily associated with the action the user wants to perform. When creating the toolbars for Microsoft Excel 3.0, we designed a simple test to allow us to decide between various icons for each function. We could generate any number of icons, test them, and decide which icon was most effective. Because we were choosing from among competing designs, we were more interested in collecting large amounts of data quickly and looking for big effects than we were in designing a very sensitive test.

We created a program in Visual Basic called Iconotester. Iconotester is a good example of a usability tool that provides various groups—in this case the product team, the visual interface design group, and the usability group—with data that address a particular design problem.

Iconotester consists of three screens. The first screen displays 15 icons and 18 short verbal descriptions of the icons. We included more descriptions than icons so that subjects could not use the process of elimination to interpret difficult icons. In addition, there was more than one icon per description. We presented the screen and asked participants to drag the description to the appropriate icon. Participants were then shown a second screen that included all of the icons. Clicking on each icon gave them a detailed description of the icon.

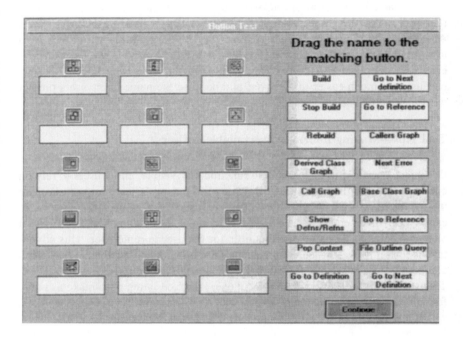

Participants were asked to spend time learning the icons' functions until they felt they knew them well.

Participants were then given a task meant to distract them for about 20 minutes. The task was part of another, unrelated, usability test involving another application. After completing the task, participants were shown the original Iconotester screen and asked to again drag the description to the appropriate icon.

Iconotester taught us that users are very poor at interpreting the functions the icons represent but are very good at recalling them. To make icons more effective we began looking for ways to make the functions associated with icons more discoverable. The Microsoft Office program management group

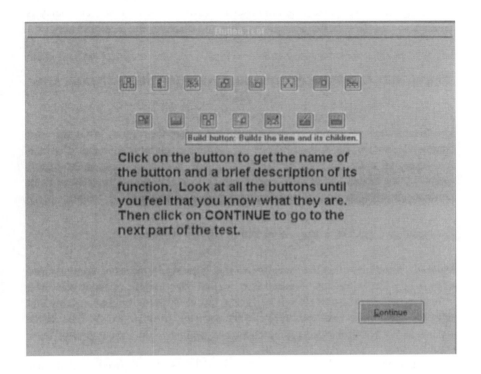

proposed ToolTips as a solution.

ToolTips are text descriptions of toolbar icons that appear when the cursor pauses over an icon. We were careful to design ToolTips in such a way that they did not interfere with normal use. They had to appear quickly enough to help users discover the meaning of the icons but not so quickly as to be intrusive. We tested ToolTips and arrived at an optimal time for the delay. Tips do not appear unless the user pauses over an icon. Tips do not accidentally or routinely appear but do appear quickly enough to make the response seem natural

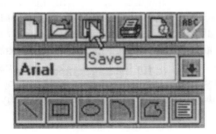

when a user is trying to determine the function of a toolbar button. ToolTips are also designed to recognize when the user is browsing the toolbar. The delay is reduced so that the user can scan the toolbar and discover the available actions.

## USING THE CONTEXT TO DETERMINE THE USER'S GOALS AND ACTIONS

It is not always necessary to make complex inferences about what the user wants or intends to do. In many cases the context is unambiguous and it is relatively easy to make the right choice for the user. While working on Microsoft Excel 4.0 we began looking for opportunities to make simple inferences from the data entered by users and anticipate user goals and actions.

### Recognizing User Intention—AutoFill in Microsoft Excel 4.0

We knew from instrumented version data that copying formulas or numbers into a range of cells is another frequent user action. The content of most cells in a typical spreadsheet is entered not by typing but by copying to create series that replicate a few cells that were typed in by the user. In other words, data series are one of the building blocks of spreadsheets, and users do a great deal of work creating them.

Our analysis of a large number of spreadsheets also showed us that they contained other sequences of cells that the user sees as data series, such as sequences of dates, times, partial dates, ordinal numbers, combinations of text and numbers and multicell blocks replicated according to some pattern (see Figure 6). There were even specialized sequences peculiar to individual companies, such as product identifications and lists of company division names. These were all series that users had to type in, because Microsoft Excel had no way of recognizing and extending these types of series.

It occurred to us that we might be able to reliably recognize the user's intended series based on the start of the series. If we could do this accurately, we could provide a single command that would not only make it easier to create the common series already possible with Copy but most of the other series as well. Another insight was that the user interface for this feature might be presented as a generalization of resizing, whereby the user would grab a handle on a cell or small range with the mouse, and "stretch" the range over a larger range. The data in the initial range would then be interpreted and extended using a pattern appropriate for its content to fill the larger range. This direct manipulation user interface for extending a series worked very well in simple cases; our goal was to develop heuristics to make it predictable for all likely cases of starting ranges.

**Table 1. Examples of Series.**

Data Type Series Example
Formula
Replicate =A1 + 1; =B1+ 1; =C1 + 1; . . .
Date Increment Jan 1, 1993; Jan 2, 1993; Jan 3, 1993. . . Jan 1993; Feb

Number Replicate 1; 1; 1; 1 . . .Text + number Increment number part Q1;
Q2; Q3 . . . Product 1; Product 2; Product 3 . . .Special text Special sequences
(The user can define custom sequences.) Jan; Feb; Mar . . . 1st; 2nd; 3rd;
4th . . . Mercury; Venus; Earth; Mars . . .
Any other text Replicate Tree; Tree; Tree . . .
1993; Mar 1993 . .

| Data Type | Series | Example |
|---|---|---|
| Formula | Replicate | =A1 + 1; =B1+ 1; =C1 + 1; . . . |
| Date | Increment | Jan 1, 1993; Jan 2, 1993; Jan 3, 1993. . . |
| | | Jan 1993; Feb 1993; Mar 1993 . . . |
| Number | Replicate | 1; 1; 1; 1 . . . |
| Text + number | Increment number part | Q1; Q2; Q3 . . . |
| | | Product 1; Product 2; Product 3 . . . |
| Special text | Special sequences | Jan; Feb; Mar . . . |
| | | 1st; 2nd; 3rd; 4th . . . |
| | (The user can define custom sequences.) | Mercury; Venus; Earth; Mars . . . |
| Any other text | Replicate | Tree; Tree; Tree . . . |

We found that we could characterize the various type of series according to three parameters: the width of the block to be replicated or extended; the height of the block; and whether the block was to be extended across, down, or in both directions simultaneously.

Although Microsoft Excel supports more advanced cases, for simplicity's sake we will discuss in detail only the case of one-dimensional series where the replication block has a width of a single cell. This includes all the series that were possible with the original Copy command as well as many other useful series. Identifying the type of series based on the first cell was relatively easy. It was also often possible to determine the increment, if any, based on only the first cell. For example, if the first cell were January 1994 we could reliably predict that the following cells should be February 1994, March 1994, and so on. Similarly, January 15, 1994, could be extended as January 16, 1994, January 17, 1994, and so on.

However, we found that the first cell was not always sufficient to allow us to reliably predict the succeeding cells. For example, if the first cell were the number 1, the succeeding cells could be 1, 1, 1 . . . , or 2, 3, 4 . . . , with both alternatives being reasonably probable. Furthermore, even in the case of a first cell being January, the user might wish to have a series that lists, for example,

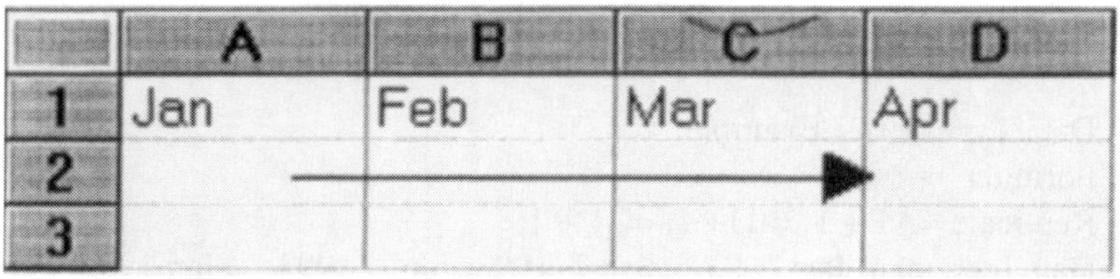

| | A | B | C | D |
|---|---|---|---|---|
| **1** | Jan | Feb | Mar | Apr |
| **2** | | | | |
| **3** | | | | |

Figure 6. One-dimensional fills extend a single series using a default increment.

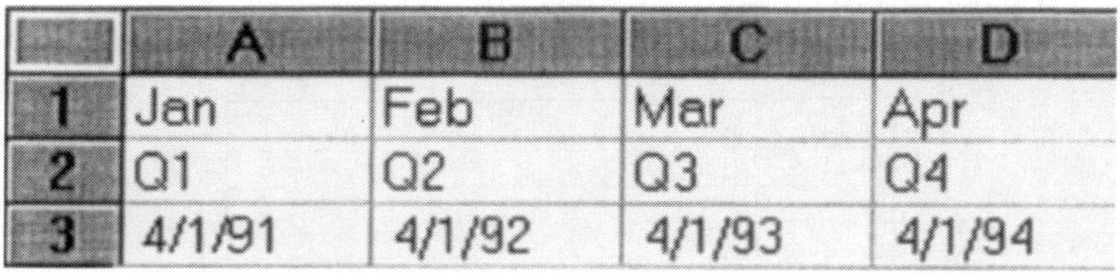

| | A | B | C | D |
|---|---|---|---|---|
| **1** | Jan | Feb | Mar | Apr |
| **2** | Q1 | Q2 | Q3 | Q4 |
| **3** | 4/1/91 | 4/1/92 | 4/1/93 | 4/1/94 |

Figure 7. A multirow one-dimensional fill simply completes parallel fills.

every second month. In the previous examples using dates, the user might occasionally want to change the default increment.

Thus, although the replication of a single cell was an important case and often all that the user needed, we had to provide a way to override the default behavior. Our decision to implement the AutoFill feature as a resizing of a smaller range into a larger range proved to support this case very well. If the user started with a range that was two cells wide, we used the difference between the values of the first and second cells to determine the remainder of the series. This worked equally well for numbers and for any of the more specialized series types, and usability testing showed it to be readily discoverable by users.

We noticed that a few users learned to enter two values when they wanted to start any series, even though in many cases the default result of filling starting with a single cell would have given the correct result. Although entering two values was not the most efficient way to complete such series, it did produce the expected result and with substantially less effort than would be required if the user typed in the entire series.

We knew that users had become so accustomed to using the Copy command that they would not switch to a new way of creating series unless it was significantly easier to use and gave predictable and reasonable results. We considered results reasonable if a user could look at the result of a fill, infer what

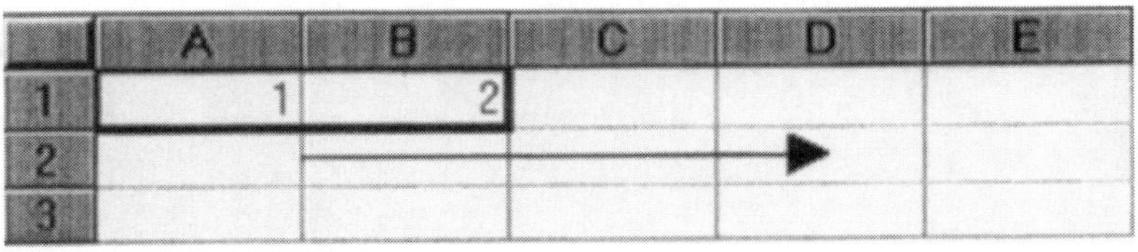

Figure 8. Extending a series using a custom increment.

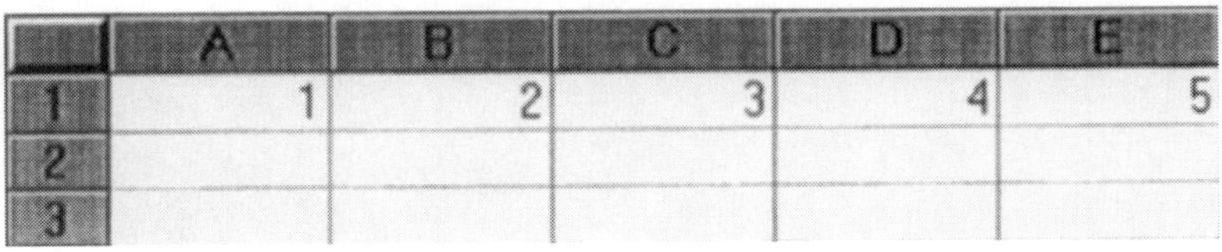

Figure 9. The series is based on the difference between the first two cells.

rules were applied, and use this knowledge to adjust the size or content of the starting range to manipulate the result.

Many of the AutoFill cases were predictable, but there were clearly cases where user expectations in a given situation might vary. There were also many conceivable cases of "garbage data" for which any prediction was difficult. However, we felt that it was essential that all cases be at least reasonable, since unexpected results could undermine a user's confidence in the entire feature. We worked hard to research all possible types of starting ranges to develop appropriate and useful, or at least reasonable, rules for extending them. Two examples illustrate this point (see Figures 10 and 11).

About three months after we shipped AutoFill in Microsoft Excel 4.0, we collected instrumented data that showed that the feature had become the 18th most frequently used command in the user interface, out of over 300 possible commands. The statistics also showed that AutoFill was used by 49% of all users. This is a very high level of use for such a new, and optional, feature. New commands rarely break into the top 20 in usage, given such common existing commands as Open, Save, Clear, and Insert.

The instrumented data also showed the following percentages of users performing fills on each of the three main types of cells:

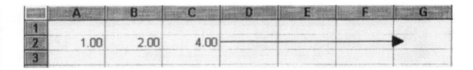

Figure 10. Extend a series based on the linear fit.

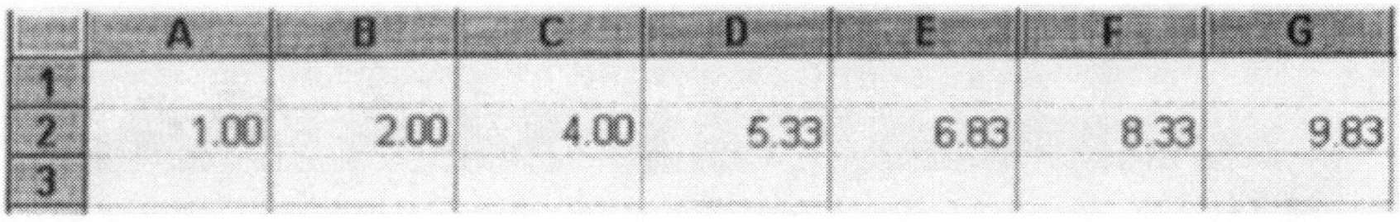

| | A | B | C | D | E | F | G |
|---|---|---|---|---|---|---|---|
| 1 | | | | | | | |
| 2 | 1.00 | 2.00 | 4.00 | 5.33 | 6.83 | 8.33 | 9.83 |
| 3 | | | | | | | |

Figure 11. The results of extending a series based on a linear fit.

- Formulas 31%
- Integers/Dates 35%
- Strings 31%

The fact that AutoFill is used frequently on a wide range of data types indicates that our heuristics are effective and that the feature was a significant addition to spreadsheet technology. The feature's success would not have been possible without a careful application of heuristics to handle the nondeterministic cases that arose from a generalization of the Copy command.

**Selecting the Correct Range in Microsoft Excel 5.0**

Automatically selecting a range in Microsoft Excel is a simple example of how it is possible to use the context to help the user. When the user selects Sort to sort a range in a spreadsheet, Microsoft Excel automatically selects the range as long as one cell in the range is selected (see Figures 14 and 15).

Selecting the correct range for an operation is very reliable, because almost all of the information that Microsoft Excel needs to determine the correct range is available. Although fairly simple, it is very useful. It prevents the very common error in Object–Verb interfaces of choosing the command without having

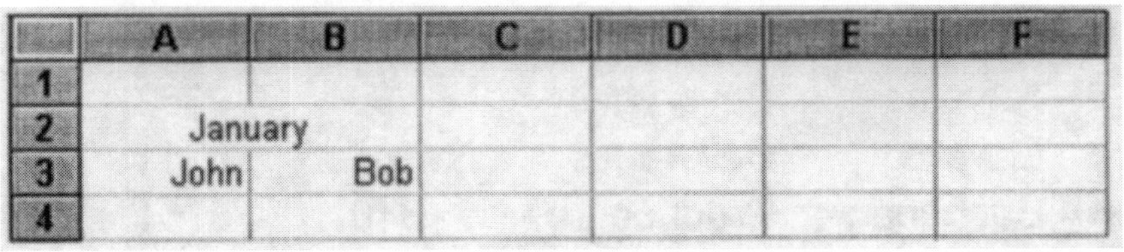

| | A | B | C | D | E | F |
|---|---|---|---|---|---|---|
| 1 | | | | | | |
| 2 | January | | | | | |
| 3 | John | Bob | | | | |
| 4 | | | | | | |

Figure 12. Extend a block of data by a combination of replication and increment-ing.

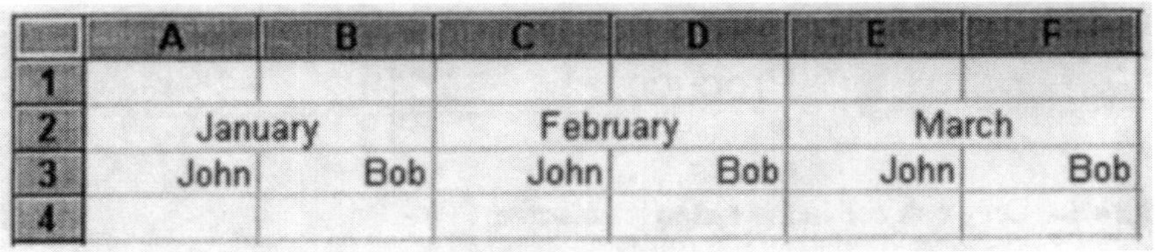

| | A | B | C | D | E | F |
|---|---|---|---|---|---|---|
| 1 | | | | | | |
| 2 | January | | February | | March | |
| 3 | John | Bob | John | Bob | John | Bob |
| 4 | | | | | | |

Figure 13. The results of extending a block of data.

chosen the correct range. It also prevents common errors such as including the column titles in the range selection and sorting them, or selecting some but not all the columns in the range and scrambling data. Although it does not work when the user wants to sort a range within a range, it works in a sufficiently large number of cases to be useful.

**Using the User's Data to Make Actions Recognizable—Sorting in Excel 5.0**

Microsoft Excel also intelligently recognizes the user's own data and uses it to improve the interface. In the earlier example Sort recognizes the headings and uses them in the Sort dialog box. Previously, users had to sort on a cell location. In the earlier example (see Figure 15) the user would have to choose a location in column B, say B7, to sort by *Type*. Microsoft Excel 5.0 recognizes the head-ings in the range and displays them in the Sort Dialog Box. Users can sort on information that they have provided and that is meaningful in the context of the task they are performing.

Microsoft Excel guesses that the first row is a header, because most tables that users create have headers. When the users sorts the data, the header row is not sorted. When there are no headers, the user can correct the error by check-ing *No Header Row*.

A related way Microsoft Excel 5.0 uses existing data to simplify a list man-

| | A | B | C | D |
|---|---|---|---|---|
| 1 | Salesperson | Type | Units | Sales |
| 2 | Suyama | Beverages | 690 | 9862 |
| 3 | Davalio | Beverages | 767 | 6711 |
| 4 | Buchanan | Produce | 340 | 8751 |
| 5 | Buchanan | Beverages | 4997 | 656 |
| 6 | Davalio | Produce | 2662 | 6715 |
| 7 | Suyama | Produce | 744 | 2656 |
| 8 | Suyama | Beverages | 2898 | 7538 |
| 9 | Buchanan | Meat | 5889 | 4953 |
| 10 | Davalio | Produce | 7549 | 1500 |

Figure 14. Only one cell in the range is selected.

| | A | B | C | D |
|---|---|---|---|---|
| 1 | Salesperson | Type | Units | Sales |
| 2 | Suyama | Beverages | 690 | 9862 |
| 3 | Davalio | Beverages | 767 | 6711 |
| 4 | Buchanan | Produce | 340 | 8751 |
| 5 | Buchanan | Beverages | 4997 | 656 |
| 6 | Davalio | Produce | 2662 | 6715 |
| 7 | Suyama | Produce | 744 | 2656 |
| 8 | Suyama | Beverages | 2898 | 7538 |
| 9 | Buchanan | Meat | 5889 | 4953 |
| 10 | Davalio | Produce | 7549 | 1500 |

Figure 15. Microsoft Excel automatically selects the correct range.

agement task is that it analyses the contents of each column in a sorted range looking for unique items. When the user chooses *Data-Filter-Autofilter* Microsoft Excel creates a dropdown list of the unique items in each column, allowing the user to pick from the list to filter based on any items in a given column.

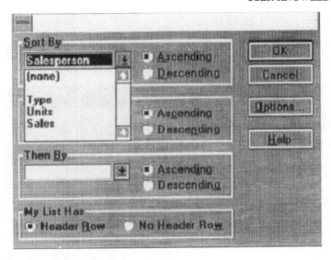

Figure 16. Microsoft Excel displays the user's headings.

| | A | B | C | D |
|---|---|---|---|---|
| 1 | Salespers⬇ | Type ⬇ | Units ⬇ | Sales ⬇ |
| 2 | Suyama | Beverages | 690 | 9862 |
| 3 | Davalio | Beverages | 767 | 6711 |
| 4 | Buchanan | Produce | 340 | 8751 |
| 5 | Buchanan | Beverages | 4997 | 656 |
| 6 | Davalio | Produce | 2662 | 6715 |
| 7 | Suyama | Produce | 744 | 2656 |
| 8 | Suyama | Beverages | 2898 | 7538 |
| 9 | Buchanan | Meat | 5889 | 4953 |
| 10 | Davilio | Produce | 7459 | 1500 |

Figure 17. MS Excel 5.0 provides dropdown lists for filtering.

**Extending Sort-Subtotals in Microsoft Excel 5.0**

Subtotals is an extension of Sort that both recognizes the users' data and allows the user to make a choice based on that data. For many users Microsoft Excel is a tool for managing lists of data. While observing users managing lists we saw that adding subtotals was a frequent goal. Once a list of data has been sorted it is easy for Microsoft Excel to add subtotals—Microsoft Excel allows the user to choose what to subtotal (see Figures 18 and 19). If users add subtotals, they may want to view the data as totals and subtotals. Microsoft Excel outlines the list so that the user, using the outline symbols, can produce a summary view (see Figure 20).

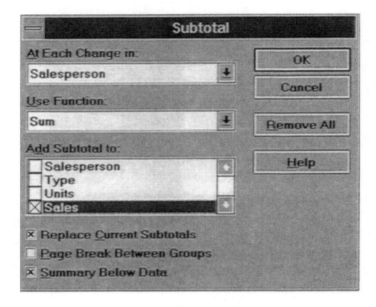

Figure 18. The users chooses the fields to subtotal.

| 1 2 3 | | A | B | C | D |
|---|---|---|---|---|---|
| | 1 | Salesperson | Type | Units | Sales |
| | 2 | Buchanan | Produce | 340 | 8751 |
| | 3 | Buchanan | Beverages | 4997 | 656 |
| | 4 | Buchanan | Meat | 5889 | 4953 |
| | 5 | Buchanan Total | | | 14360 |
| | 6 | Davalio | Beverages | 767 | 6711 |
| | 7 | Davalio | Produce | 2662 | 6715 |
| | 8 | Davalio | Produce | 7549 | 1500 |
| | 9 | Davalio Total | | | 14926 |
| | 10 | Suyama | Beverages | 690 | 9862 |
| | 11 | Suyama | Produce | 744 | 2656 |
| | 12 | Suyama | Beverages | 2898 | 7538 |
| | 13 | Suyama Total | | | 20056 |
| | 14 | Grand Total | | | 49342 |

Figure 19. Microsoft Excel then adds the appropriate subtotals.

| | A | B | C | D |
|---|---|---|---|---|
| 1 | Salesperson Type | | Units | Sales |
| 6 | Buchanan Total | | | 14360 |
| 9 | Davalio Total | | | 14926 |
| 13 | Suyama Total | | | 20056 |
| 14 | Grand Total | | | 49342 |

Figure 20. A summary view of the data.

## USING THE USER'S GOAL TO STRUCTURE A TASK

Software products can allow users to accomplish many complex tasks for which they lack the domain knowledge. The user's problem is not understanding the software, but understanding the task. Microsoft Publisher is an example of such a product. It is a desktop publishing program for nonprofessionals that is specifically designed to help users accomplish tasks for which they lack the knowledge and background. The designer's problem is to help users understand what tasks they can accomplish with the program and what information is relevant for accomplishing those tasks. Publisher's solution was the Publisher Wizard.

### The Publisher Wizard

The Publisher Wizard accomplishes four things for the user:

1. It allows the user to pick from a range of possible publishing solutions— solutions of which they may not be aware.
2. It helps users define and choose the relevant constraints.
3. It does the complex work of creating a template so that the user can create the content.
4. It allows the user to learn by watching Publisher create the final product.

For example, a user who is not a designer but needs a brochure for a mailing may not be aware of the various types of possible brochures, or how to create one. When the user starts Publisher, or creates a new file, Publisher displays the Wizard. The Wizard gives the user a number of possible solutions. When the user selects an option, the Wizard displays a visual representation so that the user can map the possible solution to their own visual representation (see Figure 22). In effect, the wizard, based on the goal the user has selected, supplies the user with the expert knowledge they need to create the solution.

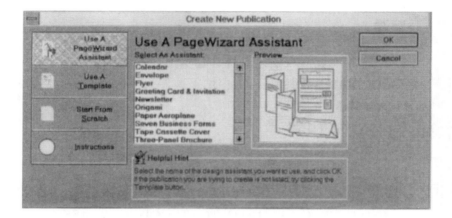

Figure 21. The opening screen of the publisher wizard.

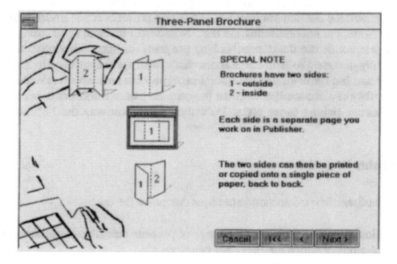

Figure 22. MS Publisher gives the user information about their choice.

Suppose the user chooses the Three-Panel Brochure from the Select An Assistant list (see Figure 22). The Wizard then provides more specific information about the solution and prompts the user for the relevant options. In this case the relevant constraints are style, fold, and whether the brochure will be handed out or mailed (see Figures 24, 25, and 26).

Publisher then creates a template of the brochure for the user and the user fills in the content (see Figure 26). The user has been able, with the help of the Wizard, to do some real work that they would not otherwise have been able to do. However, there two costs to the user: First, because the Wizard has done most of the work, the user hasn't learned anything about how to create a brochure; second, because the user does not know exactly how the brochure was created, it can be difficult for the user to modify the brochure. Thus, another goal of the wizard is to teach the user something about using Publisher to do desktop publishing. The Wizard invites the user to watch the process and learn how to customize the brochure or create another brochure (see Figure 27).

As publisher creates the brochure it tells the user which step it is performing and completes the step on the screen so that the user can follow it. The user can control the speed of the process (see Figure 29). If the Wizard is going too fast and the user cannot follow the steps, the user can slow the Wizard down. If the user wants to create the brochure more quickly, the user can speed the Wizard up.

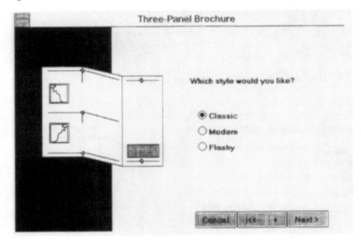

Figure 23. The users chooses a style.

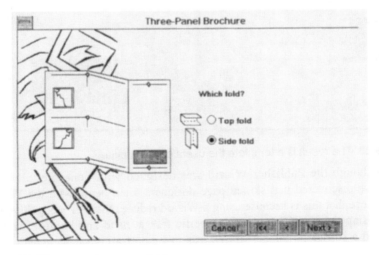

Figure 24. The user chooses a fold.

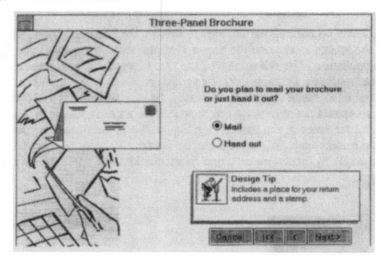

Figure 25. The user chooses a distribution method.

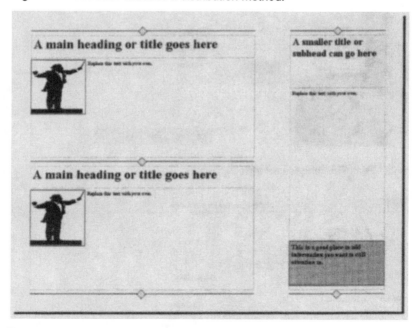

Figure 26. The result is a template the users can complete.

Although the Publisher Wizard was designed with nonprofessionals in mind, we discovered that skilled page designers often begin with the Wizard. We believe that this is because using a Wizard reduces the user's cognitive load by focusing their decisions on the specific task at hand and because we have provided a sufficient number of Wizards to support a high percentage of user projects.

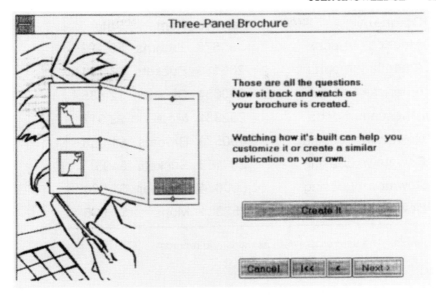

Figure 27. The user can choose to watch MS Publisher create the brochure.

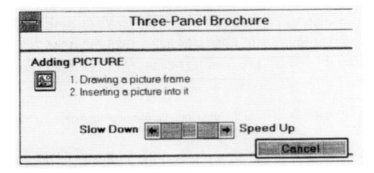

Figure 28. The user can control the speed of the presentation.

## The Pivot Table Wizard

Wizards are also useful in cases where the user may know all the operations to produce what they want, but it requires a lot of work on the user's part to create a frequent and useful result. The Pivot Table Wizard is an example. Cross tabulations are a common and very useful way to look at data. It is possible to create a cross tabulation in Microsoft Excel without using the Wizard, but it would require a very large number of operations and formulas. It would also be very difficult for the user to modify the result to look at data in different ways.

The Pivot Table Wizards lets users turn a simple Microsoft Excel database (or external data) into a pivot table—a cross tabulation that the user can manipulate and change on the screen (see Figures 29 and 30). The Wizard reads the

| Company | Part No. | Item | Sale |
|---------|----------|------|------|
| American Imports | 490578 | Brooms | $3,123 |
| American Imports | 206156 | Buckets | $3,784 |
| American Imports | 660684 | Cleaner | $2,885 |
| American Imports | 252331 | Mops | $2,917 |
| Downtown Leasing | 490578 | Brooms | $1,058 |
| Downtown Leasing | 206156 | Buckets | $495 |
| Downtown Leasing | 660684 | Cleaner | $1,326 |
| Downtown Leasing | 252331 | Mops | $1,966 |

Figure 29. The user enters data as fields and records.

| Sum of Sale | Item | | | |
|-------------|--------|---------|---------|---------|
| Company | Brooms | Buckets | Cleaner | Mops |
| American Imports | 3122.93 | 3783.614 | 2885.343 | 2916.774 |
| Downtown Leasing | 1057.986 | 494.822 | 1325.768 | 1965.792 |
| Grand Total | 4180.916 | 4278.436 | 4211.111 | 4882.566 |

Figure 30. Microsoft Excel turns the data into a crosstabulation called a Pivot Table.

row and column headers, and the user creates the pivot table by dragging the appropriate column header to the template (see Figure 31).

Unlike with the Publisher Wizard, in pivot tables the details of how Microsoft Excel created the pivot table are not important to the user; however, the ability to modify the results are critical. Users will often want to view the data in different ways. Pivot Table allows users to modify the results directly. The user can refresh the data if the original database changes, hide and show detail, add fields to the pivot table and move the fields around. For example, the user can change the fields in the table by dragging the field headings (see Figure 32).

Wizards also have another advantage. Graphical user interfaces are very good at making a lot of information visible and available to the user, and putting the user in control by allowing direct interaction with the display. However, they lose one advantage of character-based interfaces; they do not help the user through a series of subgoals that act as a solution path for the problem the user is trying to solve. Character interfaces typically guided the user toward an objective in a series of screens that required choices each of which solved part of the problem. Wizards accomplish the same thing in a graphical user interface.

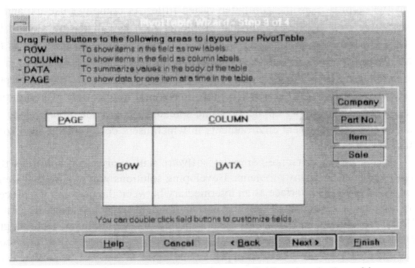

Figure 31. Users build the pivot table by dragging field names to the grid.

| Sum of Sale | Item | | | |
|---|---|---|---|---|
| Company | Brooms | Buckets | Cleaner | Mops |
| American Imports | 3122.93 | 3783.614 | 2885.343 | 2916.774 |
| Downtown Leasing | 1057.986 | 494.822 | 1325.768 | 1965.792 |
| Grand Total | 4180.916 | 4278.436 | 4211.111 | 4882.566 |

| Sum of Sale | Company | |
|---|---|---|
| Item | American Imports | Downtown Leasing |
| Brooms | 3122.93 | 1057.986 |
| Buckets | 3783.614 | 494.822 |
| Cleaner | 2885.343 | 1325.768 |
| Mops | 2916.774 | 1965.792 |
| Grand Total | 12708.661 | 4844.368 |

Figure 32. Users can modify the Pivot Table by dragging items on the screen.

## THE DIRECTION OF INTERFACE DESIGN AT MICROSOFT

In the early days of computing, computing power was applied to problems such as calculating the trajectory of artillery shells for which the inputs from the user and the outputs from the computer varied little. As computer use expanded, computers were applied to vertical problems such as calculating payroll for which the task was well structured, making it easy to map the interface to the task. Personal-computer software has been developed to solve a variety of more complex problems. For example, word processing, the most frequently performed computer task, is a much more horizontal application of computing power than an inventory application, and spreadsheets are not "applications" at all but user-development environments in which users develop their own solutions.

In the future, personal-computer software will evolve into solutions that users can apply to many problems. Developing solutions will require that we no longer think of the interface as an intermediary between the user and the processor, but as an environment that the user must understand and work in (Nardi, 1993). Developing solutions will mean bringing diverse computer technologies together to solve problems. It will also mean developing better user interfaces that give the user the power to apply technologies in new ways, so that users can attend to the problem they are trying to solve rather than to the technology they are using to solve it. Advances in the user interface will depend to some extent on advances in interface technologies, such as new pointing or input devices, three-dimensional interfaces, artificial intelligence, or virtual reality. But, more than anything else, advances in user interface design will require more knowledge of users and the environments in which users work and live. It is no longer possible to adapt a metaphor and extend it until the design is complete. Often metaphors do more to limit the size of the designer's problem than they do to help the user. Nor is it sufficient to develop a set of "interaction principles" and extend them. User-interface design will require more empirical data about users and better ways of getting that data into the design. At Microsoft we are moving toward better ways to collect and make sense of user data.

We are currently confronting such issues as the role of domain knowledge in user-interface design and how domain knowledge can help us develop intelligent agents. We are interested in how we can use the power of the computer to create adaptable user interfaces that recognize user intentions and do more of the work for the user or provide assistance at the time the user needs it. We are also looking at better ways to make the users' interactions with the computer more interesting and natural. Our goal will continue to be following an evolutionary path that allows our users to leverage the knowledge they already have rather than forcing them to take the difficult step of abandoning well-learned strategies for new and less familiar ones.

# REFERENCES

Browne, D., Toterdell, P., & Norman, M. (Eds.). (1990). *Adaptive user interfaces*. London: Academic Press.

Bennett, J., Holtzblatt, K., Jones, S., & Wixon, D. (1990). Usability engineering: Using contextual inquiry. In J. C. Chew & J. Whiteside (Eds.), *Proceedings of CHI '90: Human factors in computing systems*. New York: ACM Press.

Byrne, M. D., Wood, S. D., Sukaviriya, P. N., Foley, J. D., & Kieras, D. E. (1994). Automating interface evaluation. In B. Adelson, S. Dumais, & J. Olsen (Eds.). *Proceedings of CHI '94*. New York: ACM Press.

Card, S. K, Moran T., & Newell A. (1983). *The psychology of human computer interaction*. Hillsdale, NJ: Erlbaum.

Carroll, J. M. (1991). *Designing interaction*. Cambridge: Cambridge University Press.

de Baar, D. J, Foley, J. D., and Mullet, K. E. (1992). In *Coupling application design and user interface design. CHI '92 Conference Proceedings* (pp. 259–266). New York: ACM Press.

Edmonds, E. (1993). The future of intelligent interfaces: Not just "how?" but "what?" and "why?" *Proceedings of the 1993 International Workshop on Intelligent User Interfaces*. Orlando, FL.

Gould, J. D., & Lewis, C. (1985). Designing for usability: Key principles and what designers think. *Communications of the ACM, 28*(3), 300–311.

Hefley, W. E. (1990). Architectures for adaptable human-machine interfaces. In W. Karwowski & M. Rahmim (Eds.), *Ergonomics of hybrid automation systems II. Proceedings of the Second International Conference on Human Aspects of Advanced Manufacturing and Hybrid Automation* (pp. 575–585). Amsterdam: Elsevier Science.

Hefley, W. E., & Murray D. (1993). Intelligent user interfaces. In W. D. Gray, W. E. Helfley, & D. Murray (Eds.), *Proceedings of the 1993 International Workshop on Intelligent User Interfaces* (pp. 3–10). Orlando, FL.

Howes, A., & Payne, S. J. (1990). Display-based competence: Towards user models for menu-driven interfaces. *International Journal of Man–Machine Studies, 33*, 637–653.

Hutchins, E. L., Hollan, J. D., & Norman, D. A. (1986). Direct manipulation interfaces. In D. A. Norman & S. W. Draper (Eds.), *User centered systems design: New perspectives on human–computer interaction* (pp. 87–124). Hillsdale, NJ: Erlbaum.

Johnson, B., & Shneiderman, B. (1991). Treemaps: A space-filling apporach to the visualization of hierarchical information structures. In B. Schneiderman (Ed.), *Sparks of innovation in human–computer interaction* (pp. 309–325). Norwood, NJ: Ablex.

Mannes, S. M., & Kintsch, W. (1991). Routine computing tasks: Planning as understanding. *Cognitive Science, 15*, 305–342.

Napier, H. A., Batsell, R. R., Lane, D. M., & Guadagno, N. S. (1990). *Knowledge of command usage in a spreadsheet* Program: impact on user interface design and training.

Norman, D. A. (1986). Cognitive engineering. In D. A. Norman & S. W. Draper (Eds.), *User-centered system design: New perspectives on human–computer interaction* (pp. 31–62). Hillsdale, NJ: Erlbaum.

Norman, D. A. (1988). *The psychology of everyday things*. New York: Basic Books.

Nardi, B. A., & Zarmer, C. L. (1993). Beyond models and metaphors: Visual formalisms in user interface design. *Journal of Visual Languages and Computing, 4,* 5-33.

Nielsen, J. (1993). *Usability engineering*. San Diego: Academic Press.

Shneiderman, B. (1986). *Designing the user interface*. Reading, MA: Addison-Wesley.

Simon, H. A., & Newell, A. (1972). *Human problem solving*. Englewood Cliffs, NJ: Prentice-Hall.

Simon, H. A. (1963). *Sciences of the artificial*. Cambridge, MA: MIT Press.

Suchman, L. A. (1987). *Plans and situated actions: The problem of human-machine communication*. New York: Cambridge University Press.

Sullivan, J. W., & Tyler, S. W. (Eds.). (1991). *Intelligent user interfaces*. Reading, MA: Addison-Wesley.

Whiteside, J., Bennett J., & Holtzblatt, K. (1988). Usability engineering: Our experience and evolution. In M. Helander (Ed.), *Handbook of human computer interaction*. Amsterdam: Elsevier Science.

Wright, P. C., & Monk, A. F. (1991). A cost-effective evaluation method for use by designers. *International Journal of Man–Machines Studies, 35,* 891–912.

# Chapter 7
# Extending Usability Engineering Techniques Into the Real World: A Commentary

Thea Turner
Alison Lee
Michael E. Atwood

*NYNEX Science & Technology, Inc.*

## INTRODUCTION

Most human–computer interaction (HCI) researchers and practitioners would contend that usability engineering research has come a long way in the past 50 years. As evidence of this progress, the growing number of techniques available to an HCI practitioner can be cited. These techniques can be applied at various points in the development process and require different types of information and resources. However, before rushing to proclaim progress, it is appropriate to examine the effectiveness of these techniques in real-world applications. In this chapter, we report on our experiences with usability techniques in a real-world setting over an extended time.

For the past few years, we have been involved with a project to design and field a new workstation for telephone operators. This application is one in which speed is crucial (operator work times are measured in millions of dollars per second), where systems must operate continuously and without failure, and one in which an understanding of a broad range of knowledge is required in the design task. The project involved a fixed budget and schedule for its deployment. It depends on software under development by two outside vendors.

The human–computer aspects of this project received attention at a very early stage. Actual users of the three predecessor systems were studied in the workplace. The existing task was analyzed in detail before the new system was designed. We participated in design meetings to define the redesigned operator tasks, the system functionality and its interface. In the course of these meetings, empirical findings from the studies of the systems to be replaced were discussed and used to aid in the design of the new system. The existing data was used with a number of usability techniques to answer questions that arose in design meetings and during implementation. When the new system was deployed, operator performance was studied in detail for a year to understand effective use of the system and provide information to improve the application.

A number of real-world considerations were associated with this project. Our efforts were obstructed, sometimes severely, by constraints on dollars, time, availability of personnel, and various requirements from our organization and regulatory bodies. Because there were three computer systems to be replaced by the new system, the development effort can be viewed as a redesign. As such, the implications of task changes on an installed base of thousands of telephone operators as well as assorted other support personnel had to be considered.

An overview of the project on which this work is based is presented here. We then discuss three aspects of usability engineering where we had difficulty in the integration of usability engineering into the system development process:

- influencing the real-world system development process;
- informing system redesign efforts;
- reusing and building on usability engineering results.

We use examples from the project to illustrate these areas. Finally, we propose an approach to integrate usability into the development process, thereby enriching both usability engineering and system development.

## INTEGRATING USABILITY ENGINEERING INTO SOFTWARE DEVELOPMENT

### An Overview of the OSDI Project

The Operator Services Digital Integration (OSDI) project is a multiyear project to develop and deploy a new operator workstation to support existing operator services such as Call Completion Services (CCS) and Directory Assistance (DA) and other services as necessary. Usability engineering has had a role throughout the OSDI project. Our involvement began in 1988 with Project Ernestine (Gray, John, & Atwood, 1993), in which we examined a candidate workstation through empirical trial and through the development of GOMS-like models. When we found that the workstation would not satisfy NYNEX's requirements adequately, NYNEX made the decision to develop a new workstation rather than purchase one. The new workstation was to support multiple operator functions and such an environment did not exist at NYNEX. Directory Assistance operators (who answer when dialing 411 or 555–1212) and the Call

Completion Services operators (who answer when dialing 0) were in separate offices and used different equipment. Therefore, the next step was to create an environment that combined these function to allow us to understand the new environment and to prototype a multiple-function workstation. This allowed us to learn a great deal about the operator environment and the use of a workstation in that environment. We analyzed the operator tasks along a variety of dimensions, such as the frequency of each task type, the various ways in which the task could be performed, and the common difficulties operators have in performing these tasks. As a result, we were in a position to influence the design of the new workstation.

The OSDI project involved a variety of parties. A number of NYNEX organizations were involved, including the organizations that managed the operator work force, the facilities, and the database, as well as those involved in legal and regulatory issues. A significant part of the workstation was developed by a group within our own organization. An outside vendor provided the Directory Assistance database, an early version of the Directory Assistance functionality and the system integration. The telephone switch vendor was involved in issues related to the switch interface and communication protocol.

A committee was formed to make the necessary design decisions for the Operator Interface. This Operator Interface Team (OIT) consisted of representatives from the various user groups, including HCI specialists. It met frequently during the design and implementation phase from early 1991 until late 1992. The committee evolved as implementation issues became more prominent than design issues, and still meets from time to time, as necessary.

## USABILITY ENGINEERING IN THE OSDI PROJECT

During our involvement in the OSDI project, we applied a variety of usability techniques. In the face of real-world constraints, many techniques were difficult to apply and some did not produce sufficient benefits to warrant their cost. Our experiences show that the impact of many usability engineering methods is diminished in a complex and real-world project. In addition, certain techniques are less useful in the context of a real development project than in a laboratory setting. Some common difficulties we experienced were:

1. *an inability to provide timely answers to usability questions*—the literature rarely provided useful answers; the methods often took too long to answer questions definitively;
2. *the inaccessibility of relevant design information from predecessor systems*—we had to reconstruct and reexamine aspects of the design of predecessor systems to determine what we needed to know to redesign the functionality of the workstation;

3. *the loss of design information that is needed later on or that could be used to inform the design of subsequent efforts*—design information was not consistently captured and catalogued for use later during the system's implementation and maintenance, for related design efforts that can benefit from the work accomplished in the current design or to learn more about usability by examining the implications of design decisions on actual use.

During the project, we were able to take the collective knowledge and experiences accumulated over many years—including the usability engineering results—and apply them to the design of the new workstation. During that time, our group was a partner in the workstation development process. In the OIT committee, we were part of a participatory design team made up of stakeholders from various user groups (e.g., trainers, managers, operators, developers, facilities administrators). We had a significant and influential voice in specifying the usability and functional requirements for the workstation. We worked closely with the in-house developers in the implementation and testing of the workstation and followed the system implementation to ensure that it met the usability requirements. When the workstation was deployed, we had access to a complete office of operators and could study the situated use of the workstation.

Although usability engineering is now a part of the process of system development, all of the parties involved—including ourselves—have not understood how to integrate this contribution into the whole process. The same basic problems surfaced time and time again when new usability questions came up, it was often difficult to provide timely answers and the available literature rarely provided insights. The methods for examining the empirical evidence generally took too long or were not able to answer questions definitively. Responses to our inquiries from those in the workplace were generally conflicting and did not provide enough quantitative information to make decisions, and relevant design information from predecessor systems was rarely accessible. To determine what information was necessary to redesign the functionality of the workstation, various aspects of their designs had to be reconstructed and reexamined. The relevant information leading to design decisions was not consistently captured and catalogued for subsequent use. This information is useful for the implementation and maintenance of the system, related design efforts that could benefit from the work accomplished in the current design, or the examination of the implications of a design on actual use.

Because the project encompassed many years and grew out of existing work practices, it was difficult to integrate usability engineering into the system development process. These difficulties affected our ability to develop usable software. The current state of usability engineering is inadequate to address system redesign efforts and to effectively integrate into the real-world system-development process. In addition, it was nontrivial to reuse and build on existing usability engineering results.

## Influencing the Real-World System Development Process

The common view of the HCI community is that the traditional "waterfall" method of system development does not work. Unfortunately, this is often the way usability engineering is done in practice because its methodologies reinforce a staged approach to system development. These methodologies may be proactive—analysis before design or development—or reactive—evaluation reacting to a design, prototype, or system. We need to develop methods that encourage a tighter coupling of analysis, design, implementation, and evaluation. When these processes are entwined, interacting with each other as each proceeds, a more appropriate system development methodology will result.

Proactive methods of usability engineering are methods that are performed early in the design process, and, as such, are used to guide and inform designs by anticipating potential usability problems that may occur. The earlier in a development effort this is done, the better. This should not, however, be construed as the end of usability engineering. Problems that are unforeseen by these methods may result in unexpected problems further along in the development. When usability analysis is done up front, there is a danger that further examination of a system's usability may be deemed unnecessary. Furthermore, proactive methods provide no validation of their results, no feedback as to whether the perceived risks are valid or whether the solutions avoid the predicted usability problems successfully. Because of this, the usability engineer may fail to learn important information that may be useful or necessary in this or other designs.

Reactive methods are applied after a design, a prototype, or a system implementation, and serve largely to evaluate the result of previous stages of software development. As such, the results they provide may not be timely. Fixes to problems found at this stage are often too difficult to implement for technical or political reasons. Therefore, changes to the system are not made, are "repaired" in the documentation, or are scheduled for consideration in "a future release." Relevant information is generally not propagated back to the designers or developers.

An interactive approach to usability engineering is more appropriate to an iterative system-development process, entwining analysis and evaluation with design and implementation. Decisions arise throughout a product's development cycle that can influence its usability. Fitting usability engineering into particular stages of the product development cycle will limit its effectiveness. We believe that analysis, design, implementation, and evaluation are different facets of the same activity and provide different perspectives on the same task. As such, these activities should not be construed to be independent of each other nor should they be performed in a fixed order. The further usability engineering is from the actual design or implementation, the less impact it will have.

We tried to follow the interactive approach throughout the OSDI project; however, the different responsibilities of the development team resulted in communication difficulties. In addition, the team members often did not see each other for long periods, partly due to the distance separating them. OIT meetings were held by video conference, with infrequent face-to-face meetings in a single location. Information was summarized and transmitted to developers in various locations across the country. However, sometimes relevant information was not included in the summaries, particularly information related to the intentions behind the design decisions made by the OIT. In addition, information from the developers, such as particular constraints known to the developers, were not always transmitted back. This breakdown in communication lead to the implementation of some functions in a way that violated the original design intent.

An example can be taken from the case of the CCS scenarios produced by the OIT committee. Some examples of these scenarios are (a) arranging for a call to be billed to a third party number, (b) verifying that a telephone line is functioning properly, and (c) dialing for a customer who is having difficulty placing the call. The scenarios were detailed depictions of a representative sample of CCS tasks, specifying the customer–operator dialogue, the screen contents, the operator key actions, system actions, design rationale, and any other relevant information. They were developed using the expertise within the committee, the empirical data on operator performance and task variants, and the expected capabilities of the switch, the relevant databases, and the workstation.

Walkthroughs were used to find and correct problems in task design embodied by the scenarios. Other scenarios were added to work through variations of procedure as necessary. Relevant information was added to each scenario during the walkthroughs. The detailed document that resulted was passed on to the development team. As the developers encountered questions, the scenarios were revisited and revised as necessary or new scenarios were developed to work through design points. Thus we were able to use this document to communicate between the user community and the developers.

At the time the scenarios were put together, the specification for the protocol through which the workstation was to communicate with the telephone switch[1] was incomplete. The developers of this protocol were not forthcoming on many issues of importance to the design of the workstation. In addition, they maintained that at no time could the workstation count on the switch to provide any response. This led to a number of design changes by the OSDI workstation implementers, and meant that a number of expectations as to the normal state of affairs could not be depended on. Therefore, the workstation implementers made a number of changes to the workstation so that it did not always function as specified by the CCS scenarios. These changes were not always communicated back to the OIT committee. Although some of the decisions were still consistent with the original intent, others resulted in usability flaws. These were only

---

[1] A switch is basically a computer system that controls the telephone network.

discovered during the acceptance test[2] of the workstation, at which time the workstation failed some tests because it did not perform as specified in the CCS scenarios. At this time, we had to negotiate with the developers as to whether changes should be made to the test scripts or the code.

At any given time in a real-world project, the current design is a snapshot of the current understanding of the developers. Because new and changing information results in an evolving understanding and causes changes in the design, good communication is necessary to ensure that all concerned parties are synchronized. Usability engineering must be iterative, returning to reevaluate the design as it proceeds through conceptualization, design, implementation and deployment.

**Informing System Redesign**

As system development efforts focus on the redesign of existing systems, usability engineering methods are needed to examine actual user performance in the systems that are to be replaced.  User performance on these systems provides an important source of information for the design of replacement systems. Performance difficulties in an old system should be examined to determine improvements or enhancements for the new system. Some aspects of the old system already lead to optimal performance, and designers should ensure that the new system supports optimal behavior. In addition, designers must consider how users will make the transition to the new system. Usability engineering must support these activities to ensure that relevant lessons from the old system are not ignored or rediscovered too late.

Performance data was collected for the existing systems to provide assistance in the design of the workstation's user interface.  The data provided information about the difficulties operators had in performing their tasks. Operators across a wide range of skill levels were compared on a variety of dimensions, such as time spent on each subtask, strategies for subsequent searches, types and magnitude of errors made, methods of error recovery and the rate of initial search success. Whenever possible, the performance on the old workstation was used to predict the implications of design alternatives for the new workstation. In some cases, the analysis of performance on the existing systems indicated areas where new functionality could enhance performance.

The design of the "Extend Search" function in Directory Assistance illustrates the use of empirical data in redesign. Operators using the existing workstations were often inconsistent in the strategies applied to follow up an initial unsuccessful attempt to find a listing. Although they had been taught to broaden their queries after initial searches in which no listings were found, in practice they often did not do so. Instead of removing information from their queries,

---

[2]In this case, acceptance testing is a form of system test involving test calls to the workstation in the dstination environment.

they chose to change the search by making it more specific or by changing details, even when no listings were found that match the search specifications. Thus, the follow-up search might query the same part of the database as before, or might not include some listings that should have been considered.

To make this more concrete, suppose a customer asks for a listing for Linda Cully on Main Street in Patchogue. The operator enters "CULL" in the primary field, "LI" in the secondary field and "MA" in the Street field. The workstation defaults to Suffolk county because the operator does not specify a community. After searching the database, the screen shows a listing for Linda Cullen on Main Street in Huntington. This matches the entered specifications, but is not the one requested. The operator might now enter "PA" for Patchogue in the Community field to limit the search to listings in Patchogue. In fact, because no community was initially specified, these listings were already searched and no appropriate listing was found. The requested listing could be under Dan Cully on Main Street or it could actually be around the corner on Bay Avenue. Thus, if the operator had broadened the search by removing the "LI," the "MA," or both, other listings that might be appropriate would be returned from the database. In fact, the operator should have done this according to the accepted practices—guidelines that can be superseded by experienced operators if they have reason to believe that another strategy might be more successful. Operators frequently do not follow these guidelines and often chose strategies that are not likely to succeed.

Two things are clear from this problem. First, operators are unlikely to follow a particular practice in searching the database, and might therefore take longer to find a listing or might miss it altogether. Second, even if an operator does follow the practice to broaden the search after no appropriate listings are found, the operator must take some time to decide which piece of information should be removed and then to perform this action.

To encourage operators to remove information, the new workstation included a new function to provide a simple and fast way to broaden a search using a single action. The function, called "Extend Search," is performed by pressing a key with that label. Each time the "Extend Search" key is depressed after a search, a new search is initiated that removes some of the information from the search specification, in a specified order. This function ensures subsequent searches are systematically broadened. Operators are able to broaden their searches quickly, without formulating the specifics of a follow-up search, providing good coverage of the database. In lowering the cost of the appropriate follow-up search, the number of searches that involve changes in details should also decrease.

As part of our follow-up work on the performance of the OSDI workstation, we wanted to see if the new function was accomplishing its goal. Initially, the "Extend Search" function was rarely used. A year after deployment, operators had increased their use of "Extend Search" to 14%. An additional 10% of the time operators removed information manually, suggesting the need for follow-up training to increase their awareness of the more efficient "Extend Search" function. Operators still changed details 20% of the time. We believe that we are on the right track in our redesign, but further refinements are possible to support better searching strategies after an initial failed search. For example, some operators inappropriately generalized the function to the case of a single search field, as if they expected it would further broaden the search. Perhaps it would be beneficial to add a step that shortened the search string when "Extend Search" is performed with a single search field. We are currently examining the effectiveness of the strategies that operators use and the implications for the design of a system to search databases.

New systems are often built to replace existing ones; when that happens, it is important to know as much as possible about the old system in order to avoid repeating the same mistakes. Areas that are difficult or time-consuming must be identified so that problems are fixed or enhancements added. In areas where performance is optimal, that behavior must be identified and understood to ensure that the new system can support expert behavior. Ideally, this information should be available; unfortunately, existing work practices are rarely documented. The original intention behind the existing system's design often needs to be reconstructed. This information is valuable in producing replacement systems that are more usable than the original system.

### Reusing and Extending Usability Engineering Results

Many of the usability methodologies today are quick and economical—there are good reasons for this. It can often be easier to get approval for usability engineering if it requires little additional time and resources. However, these methods sacrifice a thorough job to achieve this economy and turnaround. The potential problems and their severity are determined by intuition and theory instead of through empirical results. Important problems may be overlooked, showing up later as decreased user productivity. The cost of fixing the problem may be much greater than the cost of more comprehensive usability engineering. For example, in the operator workstation, a problem that caused operators to take an average of one second more per call would cost NYNEX up to $10 million a year. Therefore, there are times when economy is not the overriding concern in selecting which usability engineering methods to use. The methodology should be chosen with respect to balancing costs with benefits and risks.

Usability engineering should be viewed as an investment in the system, yielding a payback in its future performance. Long-term gains are possible with more informative usability-engineering methods. When simpler methods are used, the short-term gains may be offset by the costliness of problems missed, poor design solutions and the loss of information to optimize user performance in the system. It should be noted that over the long term, the cost of good usability engineering can more than recoup its costs through savings in productivity and system effectiveness. We need to balance costs against benefits, upfront costs against down-stream costs, the short-term benefits of a quick approach against the long-term benefits of a thoughtful one.

To make this cost-benefit issue more concrete, we present two recurring problems encountered in the OSDI project: One was the need to reuse usability engineering results and the other was the need to extend on, rather than rediscover, past usability engineering efforts. We are often dismayed at the lack of applicable information available in the literature. One executive asked if we could "just look it up" rather than testing the workstation with operators. Although we recognize that each system will contain elements that are unique, some of our questions could have been answered by relevant basic research on human–computer interaction. Unfortunately we were rarely able to find useful information. The results of usability engineering should contribute to a corpus of reusable or generalizable data and metrics that others can apply easily. This allows the field of HCI to grow by building on prior failures and successes. The goals of usability engineering should not simply be to evaluate yet another design, but to enable progress in system usability. Usability engineering must be an integral part of the system development process, playing an active role throughout, and producing results that can be generalized.

Several problems in which we could have used some guidance from past results were related to the use of abbreviations and the implications of uniquely identifying a specific entity from an abbreviation that may not be unique. A subtask of the new workstation involved the specification of the city or town for the requested listing. The operator types in a short code. If more than one match for the code is possible, a menu of matches is displayed. The operator then selects the desired locality. How should these codes be assigned? What would the ramifications of an intermittent menu be?

First, we should explain why the workstation required the locality to be specified precisely. Directory Assistance listings must be divided by regions in order to fit them into computer databases. These regions may be one or more counties, large cities or parts of cities, or a suburban region. Because an operator typically handles many such regions, operators must specify the region/database to be searched. In that these are large databases, containing hundreds of thousands of listings, the database vendor for the new workstation required that the locality (generally the city or town) of the listing be specified for a search. (Of course, we are all familiar with the opening "What city, please?" when we call Directory Assistance!) By specifying the locality, two benefits were postulated:

- The database search would be more efficient and thus faster, because it would only have to look at the small subset of the listings in the database.
- Fewer listings would be retrieved, thus reducing the time necessary for the operator to scan the listings after a search.

Both of these were expected to result in faster average work times per call.

The previous system used in New York required the operator to specify the region of the search, but the exact city or town (referred to as a locality) was optional. If an operator entered a locality, only the first two characters of its name were typed. Listings in any locality beginning with those two characters and matching the query details would be retrieved. If none matched, listings matching the rest of the query information but from other localities would be retrieved. In New England, the operators always specified a locality, using the first three characters of the locality. Listings would be retrieved from any locality in the region matching those three letters. Thus, the new system would use locality information in a manner that is different from both of the prior systems.

The locality in the new workstation was to be specified using a three or four letter code, depending on the region. The algorithm for the locality code was specified by the database vendor and was different for localities with one, two or three or more words. If a code matched more than one locality in the area code, a menu of possibilities would be brought up. The operator would then choose the intended locality by typing the alpha character to the left of the correct locality. After the locality was entered, the cursor moved automatically to the next field.

Discussions in the OIT committee centered on the tradeoff between three- and four-character locality codes. Each keystroke takes on average 1/3 of a second (Card, Moran, & Newell, 1983; also from our empirical data). The time necessary to select from a menu was unknown. We hypothesized that operators would quickly learn the codes, because they would use them often, and that the cognitive code for a locality would include the character associated with the menu selection. If so, the time for three-character codes without a menu would approach 1 second, and with a menu it would approach 1 1/3 second. We examined the localities for one area (Nassau and Suffolk counties of Long Island, NY) and estimated that roughly 30% of the locality requests would require menus for three-character codes and 19% for four-character codes. In both cases the largest menus would have three items, with two item menus as the average size. There would be 101 items on 48 menus in the first case, 64 items on 31 menus in the second. Menu selection would have to take over 2.7 seconds before it would be better to use a four-character code over a three-character code. Therefore, we chose to use three-character codes initially and to reexamine the issue later when more information was available.

The use of locality information was a modification of the existing operator task by the database vendor. We were certain that there would be usability problems that required refinement because the formation of the locality code was not straightforward. When the initial version of the OSDI workstation was deployed, we collected empirical data on operator performance. Because we had some concerns about how well operators would be able to use the locality algorithm, we targeted this subtask for further study in order to fine-tune the locality subtask.

We discovered that the locality subtask added a number of seconds to the work time of operators. When we looked further, we found that only a few operators learned the efficient code plus menu-selection strategy for the localities with menus. To make matters worse, operators were not always looking at the screen and thus missed the fact that they had a menu to which they must respond. These errors compounded to produce very long work times on calls involving menus. Finally, we noted that the problems with the menus increased for low-frequency localities, for items farther down a menu, and for items on longer menus. Figure 1 shows the time taken to enter each character for the three algorithms. The first character of one-character localities took significantly longer than for two words, which was in turn longer than for three-word localities. The second character of the code was significantly quicker for the one-word algorithm than the other two algorithms, which did not differ from each other. The last character took significantly longer in the algorithm for three-word localities than in the other two algorithms.

Another problem with the design of this subtask was the failure to design for error recovery and for the environment in which the system was to be used. To illustrate, we observed quite often that operators did not realize that a locality menu would appear and would proceed to key in the string for the next field. Consequently, either the operator would happen to key in a character in the next string that would be a valid selection from the menu (although not necessarily the one intended) or the characters would be considered invalid. In either case, the system gobbled up the characters intended for the next field. The system indicated an invalid menu selection by ringing an audible bell. This bell and the locality menu on the screen provided the only cues to the operator that their inputs were invalid. The audible bell would typically not be heard because it came out of the back of the workstation and was more likely to be heard by the operator sitting across from the operator who made the error. Furthermore, operators wore headsets and were often in the middle of a conversation with the customer. The best-case consequence of the error would be that the operator lost a few characters of the next string and had to re-key the string after making the right selection from the menu. The worst-case consequence of this error would be that the operator did not realize an error had occurred and initiated a search with an incorrect search string and an incorrect locality. Failure to understand the environment within which the workstation operates resulted in a subtask design that was not usable and did not allow a graceful recovery from errors. The difficulty of the locality-coding algorithms and the problematic design of the locality subtask led to long work times for the call, and at times resulted in a

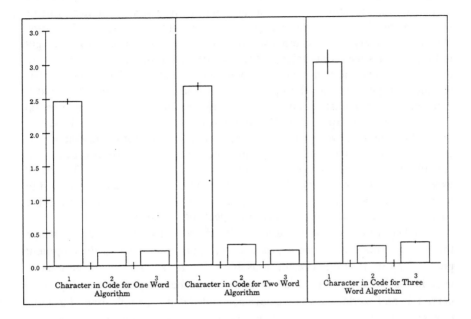

Figure 1. Time to enter each character of the locality code for the three algorithms (with Standard Error bars).

failure to successfully locate the desired listing.

Experience with real users after the designed system is deployed is important and should be a part of usability engineering methods if we are to truly improve the designs of software systems. If we had not examined the degree to which the implementation achieved the design intent, we would not have known that we failed to meet our objectives. We did not evaluate the high-level design effectively in its ability to achieve the design intention. Part of the reason is the paucity of information in the literature for guiding an investigation of the basic issues underlying the design of the locality task.

## FOCUSING THE DEVELOPMENT PROCESS ON DESIGN INTENT

In software development, it is a common practice to focus on a description of a system's own operations rather than on how the system will interact with its environment. This inappropriate focus leads developers to spend their time almost exclusively on the internal workings of the system. In fact, a key insight in Project Ernestine (Gray, John, & Atwood, 1993), our initial involvement with this project, was that there are times when the operator–customer interactions are more important than the operator–workstation interactions. This work explained why a workstation optimized for the operator–workstation interaction did not function as expected in the environment. We believe that this is part of the problem in the development of usable systems and that it is more appropriate to focus on an understanding of how the system should be constructed to support its users and to interact with its environment as a whole. The three examples drawn from the OSDI project illustrate how this understanding led to usable designs and where overlooking this understanding led to unexpected usability problems and a need to rework aspects of the system.

We believe that to produce usable software, we must fundamentally change the focus of the development process. An appropriate system development process[3] by itself will not result in usable software. As a result of our experience in the OSDI project and an examination of the practice of usability engineering in real-world settings, we have concluded that the application of usability engineering by itself is also not sufficient to develop usable systems. Rather, problems result from breakdowns in the communication of information about the system and, more importantly, about the environment into which the system is to be placed.

To change the focus of software design to the interaction of the system in its environment and to avoid communication breakdowns that prevent effective use of this information, it is extremely important to compile, maintain, and provide access to relevant information. Furthermore, this need extends beyond a mere repository of descriptive information about the design. It must include a system's purpose—the intentions beyond the design—along with constraints, implementation specifications, and other information about how a system fits into and interacts with the environment in which it is placed. It includes a history of the design and implementation, giving information about what was consid-

---

[3]A variety of system-development processes have been suggested in addition to the waterfall method, for example, "participatory design" (Ehn, 1989; Greenbaum & Kyng, 1991; Muller, 1992), "user-centered system design" (CACM, 1993; Norman & Draper, 1986), "iterative, evaluation-centered design" (Hix & Hartson, 1993), and "spiral mode of software development" (Boehm, 1988). We believe that it is important to have one, but that the particular one chosen will not be the determining factor in the usability of a system.

ered, what was rejected and why, and what was implemented, including any difficulties or dependencies.

We refer to this accumulated compilation of design information as a system's *Design Intent*. The Design Intent should be a living record of a project, accessible to all relevant people. Indeed, we argue that the Design Intent provides a platform on which to construct, analyze and evaluate, and communicate an understanding of the intended use of the system in its environment. This compilation of relevant information about the system design must begin before development gets underway, and should evolve throughout the development project. After initial deployment, the Design Intent should be available to maintenance activities or for related development efforts.

## Design Intent

In large and complex development efforts, like the OSDI project, application domain knowledge is spread unevenly across all stakeholders, including developers, maintainers, and various user groups (Curtis, Krasner, & Iscoe, 1988). Much of this knowledge is acquired, cultivated, understood and shared in the course of the development process. It is difficult to draw clear boundaries around necessary application domain knowledge for two reasons: First, this knowledge is much broader than it may initially appear; second, little research has been done to catalog the kinds of knowledge needed and used in system development beyond specifications related to system requirements and operations. Some candidates for application domain knowledge include design alternatives, evolution of designs, and design rationale. Another candidate would be "design constraints" related to the environment within which systems must operate. Because applications fit into a larger environment of work practices, it is necessary to understand this extended environment as well.

Consider the case of designing a new workstation for telephone operators. The necessary application knowledge goes beyond just the operator and the workstation. For example, constraints are placed on its use by the Directory Assistance customer. Because the customer determines much of what the operator does (Gray, John, & Atwood, 1993; Lawrence, Atwood, & Dews, 1994), it is necessary to include the knowledge of the customer in the application domain knowledge. The workstation must allow the operator to move easily to different fields (e.g., name, locality, street) in the query to follow along with the customer's order in the request.

Other constraints are placed on the workstation design by the office environment. The operators' supervisors have a stake in the design of a new workstation. Because their salaries are based on the performance of the operators in their office, the workstation must be designed so that they can effectively monitor individual and office performance. The office managers have the responsibility to train the operators when retraining is needed or when new functions are added to the operators' responsibilities (i.e., training constraints). The workstation must provide the necessary data to identify what retraining should be given.

Constraints on the workstation design are also drawn from NYNEX's business environment. In a very real sense, the operator services organizations have ownership over this effort because the workstation affects the revenues that they generate by providing operator services. The topic of monitoring extends the necessary application domain knowledge to include union-contract and state and federal legislative constraints on employee-performance monitoring. Also, vendors who provide the telephone switches and the databases add additional constraints in terms of how their systems interact with the workstation. Similarly, the public utility commissions of the various states in which NYNEX provides service impose regulatory constraints. They set rates, monitor the cost and performance of operator services, and have the power to levy fines when costs are high, expenses are deemed unreasonable, or when the level of service falls below a regulated standard of service.

All of these various constraints must be understood in the course of the development effort. Because this knowledge is spread across people in many locations and organizations, managing this knowledge is complex. The need to keep track of information in a dynamic, changing business climate poses great risk of information loss. A Design Intent would allow this information to be compiled and maintained to ensure that the knowledge was available and correct during the development effort.

## Construction

Design intent, as a vehicle to support the system development process, must allow such application domain knowledge to be recorded and constructed. Such application domain knowledge is acquired in the process of understanding the problem. Problem understanding is not an activity that can be completed before the entire system development can be completed. Because the environments into which systems are to be placed continue to evolve, problem understanding must proceed continuously and simultaneously with other system development activities. New requirements are likely to emerge as understanding matures. Existing requirements may evolve as dependencies, constraints, and specifications emerge and are more clearly understood. Thus, one characteristic of design intent is that it is provides a means to support the construction of this evolving problem understanding.

## Analysis and Evaluation

The analysis and evaluation of the design intent is clearly a necessary aspect of system development. Not all aspects of a system design can be resolved during its design and implementation. Sometimes it is necessary to conduct an active and ongoing analysis and evaluation of the system in use to detect situations in which the actual use runs counter to the expected use. Again, this focus on having an active, rather than passive, understanding of the system is an uncommon practice, but as we discussed earlier, it is necessary to the development of usable systems.

When systems are developed in an evolutionary manner, this evolution is generally not well documented. This is partly because evolution tends to be guided by failure, and failure is rarely documented. As Petroski (1992) argued, it is not so much the case that "form follows function" as that "form follows failure." Alexander (1964) went farther, stating that "we should always see [design] as a negative process of neutralizing the incongruities, or irritants, or forces, which cause misfit." Both the "Extend Search" and locality examples reveal the value of such analysis and evaluations. The usability problems we identified and the characterizations of the problems are valuable insights. They have provided us with information for enhancing the design. Furthermore, the knowledge relevant to actual use of both system features provides a corpus of information about real-world use of two fairly common functions performed in computing systems (i.e., query formulation and abbreviation generation). Thus, it is important to examine misfits and retain the findings not only to further our understanding of the problem, but to create a persistent repository for this information so that future redesigns or future explorations of similar issues may proceed by reusing and extending from these findings.

## Communication

Many people are required to represent all the necessary application domain knowledge, even on simple applications. Omitting any of this knowledge, through a lack of communication or simple oversight, jeopardizes the success of the development effort. It is naive to expect that a single individual, or a small group of individuals, can accumulate all relevant application knowledge. Because many kinds of expertise are required, teamwork is essential. To avoid the "Tower of Babel" problem (Brooks, 1975), communication between a diverse group of people is required.

Similarly, communication bottlenecks and breakdowns do not typically occur because one group neglects to communicate with another; rather, they occur because one group does not recognize that they should communicate with another. That is, someone does not know that they have knowledge that another requires, and the other person typically does not know what information they are missing. The underlying problem is that many people are required to develop a system, but each has their own view of it. In addition, each does not recognize that others have different views of the system and cannot communicate effectively with many of the others. Our solution to this "symmetry of ignorance" problem (Rittel, 1984) is to provide a common artifact—design intent—through which many people can communicate more effectively.

## CONCLUSION

A development effort generates a great deal of information. The information may be descriptive or functional, may be stored permanently or temporarily, and may exist in machine-readable or human-readable forms. When the development process works well, an evolving understanding of this information is constructed. The software design, its documentation, the training materials, and other artifacts that must be generated are derived from this understanding. Design intent enables developers to construct, analyze and evaluate, and communicate an evolving understanding of how the system will fit in and interact with the environment into which it will be placed.

## REFERENCES

Alexander, C. (1964). *Notes on the synthesis of form*. Boston: Harvard University Press.

Brooks, F. P. (1975). *The mythical man-month*. Reading, MA: Addison-Wesley.

Boehm, B. (1988). A spiral model of software development and enhancement. *IEEE Computer, 21*(25), 61–72.

CACM. (1993). Special issue on human-centered design. *Communications of the ACM, 36*(6).

Card, S. K., Moran, T. P., & Newell, A. (1983). *The psychology of human-computer interaction*. Hillsdale, NJ.: Erlbaum.

Curtis, B., Krasner, H., & Iscoe, N. (1988). A field study of the software design process for large systems. *Communications of the ACM, 31*, 1268–1287.

Ehn, P. (1989). *Work-oriented design of computer artifacts*. Hillsdale, NJ: Erlbaum.

Gray, W., John, B., & Atwood, M. (1993). Project Ernestine: Validating a GOMS analysis for predicting and explaining real-world task performance. In *Human-computer interaction* (Vol. 8, pp. 237–309). Hillsdale, NJ: Erlbaum.

Greenbaum, J., & Kyng, M. (1991). *Design at work: Cooperative design of computer systems*. Hillsdale, NJ: Erlbaum.

Hix, D., & Hartson, H. R. (1993). *Developing user interfaces: Ensuring usability through product and process*. New York: Wiley.

Lawrence, D., Atwood, M. E., & Dews, S. (1994). Surrogate users: Mediating between social and technical interaction. In B. Adelson, S. Dumais, & J. Olson (Eds.), *Proceedings of CHI'94* (pp. 399–404). Reading, MA: Addison-Wesley.

Muller, M. J. (1992). Retrospective on a year of participatory design using the PICTIVE technique. In P. Bauersfeld, J. Bennett, & G. Lynch (Eds.), *Proceedings of CHI'92* (pp. 455–462). Reading, MA: Addison-Wesley..

Norman, D., & Draper, S. W. (Eds.). (1986). *User-centered system design: New perspec-*

*tives on human-computer interaction.* Hillsdale, NJ: Erlbaum.

Petroski, H. (1992). *The evolution of useful things.* New York: Knopf.

Rittel, H. (1984). Second-generation design methods. In N. Cross (Ed.), *Developments in design methodology* (pp. 317–327). New York: Wiley.

# Chapter 8
# Usability Engineering
# in the Year 2020

**John C. Thomas**

*Human–Computer Interaction*
*NYNEX Science and Technology*

## INTRODUCTION

The human species is arguably most differentiated by our explicit and self-conscious preoccupation with the past and the future. We have learned to leave records that can be communicated across generations. Looking to the past, we may learn lessons that are applicable to the present. Looking to the future, we may guide our actions to produce a particular kind of future or, at least, prepare for coming events. In this chapter, a largely optimistic view of usability engineering is presented. By imagining what the world of HCI will be like in 2020, a number of issues are raised about what the desired state of our field should be. To the extent that consensus exists around any of these issues, such an envisionment serves a heuristic guide to help bring about any agreed-on future more quickly. To the extent that certain aspects of the envisioned future are deemed undesirable, we maximize our chances of avoiding undesirable outcome by acting now. It is also hoped that an examination of where we may be headed will shed light on our current actions and decisions.

The reader is no doubt familiar with trends such as those shown in Figures 1 and 2 indicating that both computation and communication are greatly increasing in capacity. In addition to technical capacity, however, an equally important trend is the ubiquity of these technologies. For illustration, I have graphed the number of separate and distinct interfaces to electronic devices that I have had to learn at various times in my life (see Figure 3).

If the capacities illustrated in the first two figures are going to impact human beings positively and not just generate confusion at a faster rate, then the field of HCI will be called on to deal with the third trend; namely, how can we provide increased functionality to users with a minimum of unnecessary and

# Raw Compute Capability

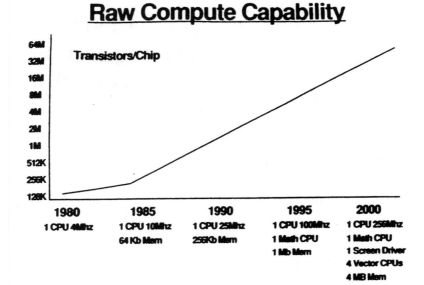

Figure 1.  Raw computer capability.

## Fiber Performance Trends

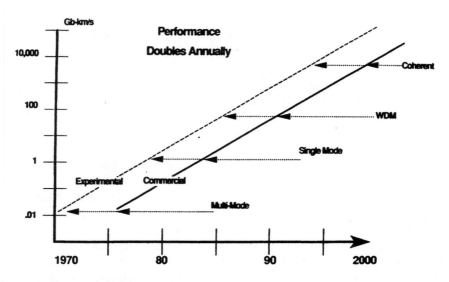

Figure 2. Fiber performance trends.
Figure 2. Fiber performance trends.

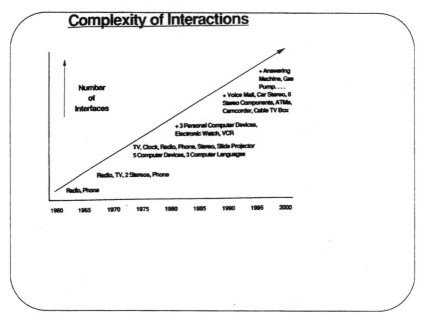

Figure 3. Complexity of interactions.

arbitrary complexity? HCI itself will have to adapt and change tremendously over the next quarter century. If the field succeeds in dealing with the challenges outlined below, the surprising conclusion of this chapter is that HCI itself will be nearly ubiquitous!

In this chapter, the future of HCI will be examined from several perspectives. First, I discuss the context in which HCI may operate. Second, I explore the impacts on methodology. Third, I attempt to explicate the overarching metamethodological issues. Fourth, I hypothesize major technological changes which will affect our work. Fifth, I take a look at a surprising conclusion dealing with the scope of problems addressed by HCI.

## CONTEXT

Large institutions have arisen with written history; therefore, large institutions are so (tautologically) much a part of the written history and current context that defines our awareness that we may well take it as axiomatic that they will exist forever. However, large human institutions did not exist for most of humankind's existence on earth and it may well be that they will not exist forever either. If any of the following types of institutions no longer exist in 2020—governments, universities, or large corporations—then clearly, the political, economic, and social context within which we in HCI work will be vastly different than it is today.

At least four separate scenarios vis-à-vis large institutions seem reasonable. First, we may find that the increase in bandwidth, storage, and computational power increases centralization (that is, power wealth, and information may be largely controlled by a few very big institutions). Second, it may be that these same trends lead to radical decentralization. Third, although the way in which these institutions function may change, the institutions themselves may play a role not unlike that they enjoy today. Fourth, there may be a mixture of all three scenarios mentioned here earlier.

## Increased Centralization

*Rationale for increased centralization.* Because the political, economic, and social context in which we do our work has such a profound effect on what and how we operate, it is worthwhile to examine the various scenarios separately, in terms of likelihood as well as implications. The first, and personally most distasteful, scenario would argue that the explosion of compute/communicate technology (referred to as C/C technology for brevity) would lead to an increase in centralization. Several new and important emerging technologies depend on big as a facilitating property. For instance, the efficacy of speech recognition technology, machine vision, natural language processing, and efficient video encoding are all probably monotonic with the number of examples. This implies that a central agent (government? communications company? computer company? union?) will have a relative advantage over many small entities attempting to develop these technologies. In addition, financial power may lead to the ability to utilize more computational power on these pattern recognition problems, also thus leading to better performance. Finally, larger institutions will have the financial resources to attract a critical mass of top-notch scientists and engineers to work on such challenging technologies (although there are many examples of "breakthrough" technologies that were created by small companies despite these apparent disadvantages; in addition, consortia and government agencies such as NIST can create "critical mass" in other ways).

*Centralized and decentralized communication technologies.* In order to see the interplay between the forces of centralization and decentralization, it may be instructive to consider the contrast between the video/TV/movie business and the phone business (at least up until the recent past). Access to the phone network has been relatively cheap and the network itself is an "anyone-to-anyone" network. A millionaire may have a fancier looking phone, but, historically, a millionaire's ability to influence someone over the phone depended on their own voice and their own ability to generate content. By contrast, TV and movies have been very expensive to produce. The ability of an individual to influence others via these media depends not only on their own abilities but on an army of technicians, make-up artists, writers, producers, directors, and other specialists. TV is "broadcast" and movies are distributed—the communication is distinctly unidirectional. A first step in any coup is to control the TV and radio stations. Significantly, and by design, former communist regimes and other dictatorships have had very little penetration of phones relative to TVs, whereas democracies in developed countries have typically had nearly universal phone access.

The difference among these communication media is further complicated by the fact that a phone network has required a huge capital investment and until recently, this huge capital investment, along with the requirements for ensuring that the interconnection "works" has led to a highly centralized phone network. Now, of course, with the breakup of AT&T in the United States and with similar opening of the phone system to competition elsewhere in the world, this picture is changing. Whereas historically, phone systems had required centralization of equipment, the content of messages was unrestricted. In democracies, this was partly for philosophical reasons, but also because no technology existed for automatically dealing with content. To be effective, "censorship" of phone conversations would have required a huge labor force listening in to many phone conversations. (Speech technology has now reached a point where this may no longer be the case.)

In contrast, although producing, advertising, and distributing a film is an expensive proposition, the actual film-making equipment is relatively inexpensive. It is possible for one private citizen to own the equipment to make a film or video tape. What recent technological advances in C/C technologies may mean is that the Internet and the phone system may soon provide an alternative and low-cost means of producing, advertising, and distributing video and eventually film-quality materials without relying on huge corporations. In order for any of this to become a practical reality, however, considerable expertise in ergonomics must be brought to bear or the results will be a chaos to potential communicators. It may now be somewhat confusing to use the "Plain Old Telephone System" in another country, but this will be nothing compared to attempting multimedia communications with various incompatible formats, devices, and user interfaces.

These patterns of difference in expense and control may be changing with new technology, however. On the one hand, TV and movies, as stated, may become relatively less expensive to produce. On the other hand, "phone" systems are beginning to offer a greater variety of bandwidths and features. It might be the case some day that a very rich or powerful person with little personal appeal in voice quality or conversational content could have their "appearance" supplemented with voice quality as well as ghost "writers" (i.e., intelligent agents).

*Impacts on the context of HCI.* Recent news seems to suggest that communication transport companies and content providers may be coming together in many cases—this may have an interesting impact on HCI work. The HCI specialists have typically worked in a scientific/technical tradition and focused on form and productivity. The "psychological" aspect of the entertainment industry, in contrast, has followed a more artistic tradition focused on content and emotional impact. In a future of "infotainment," "Infomercials," and "edutainment," it will be interesting to see how these two quite different cultures interrelate. Optimistically, the result could be both a broader fundamental scientific understanding of human psychology and richer, more rewarding, more productive experiences in the contexts of work, entertainment, and education. Historically, HCI specialists have been involved with phone companies and computer companies, but not with TV or movie companies.

One could imagine a scenario in which a large, multinational C/C company might provide content as well as capacity. Such a company might have a large staff of HCI professionals whose assignments would variously include maximizing internal operational efficiency, ensuring ease-of-use for interactive multimedia business and consumer products, and being heavily involved in the development and testing of content. In order to be effective, HCI (if that term still makes the most sense) must learn from the artistic, intuitive traditions under such a scenario. There is also something vaguely antiartistic about the notion of "user testing" or using "formal theory" to determine whether a movie has the "intended" effect. We certainly do not insist that a large, complex piece of software be designed by the untested intuitive vision of a single individual (indeed, that is what HCI specialists often have to guard against!); but will the same standard come to be applied to expressive content?

If we suppose that HCI evolved to deal effectively with emotions, motivation, and content, there might be other implications for the context in which HCI specialists worked. Whereas companies such as the hypothetical one outlined earlier would presumably cooperate on many user interface standards, there might also be an atmosphere of tight security and proprietary secrets not only for new products and services but even with respect to efficient methodologies. In the same way that governments jealously protect secret weapons, might such companies come to protect their secret methods for addressing user interface issues? We might imagine HCI in such circumstances enjoying a higher degree of respect along with the advantages of fully adequate capitalization but at the price of less cooperation and communication among companies. Perhaps in such a context, both universities (who, after all, are at once a major source of innovation and strongly prefer to publish) and government agencies (who may fund research with the proviso that it become public property) may serve an important ameliorative role. To the extent that the artistic culture drives such concerns, we might also expect an expressive pressure from the creative talent to want to share how things are created. Although predicting precisely how these various trends would impact HCI workers is difficult, we can predict that such a world would be quite different from that of today.

## Radical Decentralization

*Rationale for decentralization*. Heavier centralization is only one possible scenario. There are also factors that might lead one to consider the converse option for the future: that greater access to communication and computing power might lead to decentralization. The extreme point here might be a world in which every individual has a profile that specifies her or his abilities, interests, acceptable inputs, costs, and potential benefits/outputs. Large institutions may be a temporary anomaly in the overall history of humankind. If every person can access education, information, power, and the ability to communicate effectively with every other individual, it might be the case that large institutions have nearly outlived their usefulness. "Jobs" would be negotiated on an ad hoc coalitional basis. Whereas the duplication of physical capital (e.g., buildings, equipment) requires extensive energy and the use of limited physical resources, the unlimited duplication of information can become very cheap. Large American corporations today are often engaged in "downsizing," "rightsizing," and "outsourcing." The explosive growth of America Online, Compuserve, and the Internet (e.g., Internet users increased over a thousandfold from 1986 to 1994 according to Krol, 1994) suggests that people want to communicate and collaborate in small teams, and some of the trends in corporate America seem to dovetail. The extreme limit of today's trends to reduce "middle management" may be a "corporation" that consists of a CEO whose "workforce" is defined by an ever-changing set of agreements and arrangements carried out over the Internet.

A complicating factor in this picture is the attempt by capitalist institutions to extend the concept of physical property metaphorically to so-called "intellectual property." Experimental psychology from the 1930s experiments of N. R. F. Maier to the present have demonstrated repeatedly that our ability to attribute credit correctly to the source of a problem solving breakthrough is very error-prone. Our behavior, including our creative behavior, is influenced ubiquitously. A more narrow look at the history of science reconfirms that the notion of breaking up the ongoing web of social interaction and discovery into discrete inventions, discoveries, and ideas does extreme violence to the more complex truth. This issue becomes important in understanding whether we will see greater centralization or decentralization of work because the artificial extension of the concept of "thing"—"separate entity" to the intellectual domain, provides a leverage point for large centralized entities who have the resources to manage, direct, and trade "intellectual property."

Further, it is not just industry that may experience a massive change with the explosion of C/C technologies. Universities, for instance, might cease to exist. There are several trends in this direction in contemporary U.S. culture. For instance, much university research is now sponsored by industry; many of the student hours are taught by adjunct faculty; education and "infotainment" are available in the home; many students are reenrolling or attend class part time; there is an increasing emphasis on preparing students for the real world (as opposed to a purely intellectual pursuit). Much of today's teaching takes place via satellite, software, or network. The end result of such activities may well be a world in which individuals construct "courses" or "conversations" via a broadband network and bricks and mortar become a small part of education.

*Implications for context.* Rather than having some HCI investigators "in" academia and some "in" industry, people could bid/opt for various combinations of teaching, writing, speaking, mentoring, researching, consulting, programming and a variety of other activities. Some individuals might find their lives most productive and enjoyable if they alternate among these activities every few minutes, whereas others may find alternating on a scale of decades more rewarding.

Similar comments can be made concerning the possible decentralization of government. Significantly, the Clinton administration, however limited their implementation, has moved the White House a quantum step forward into the information age. In essence, the difference between the Greek democracy and the U.S. republic can be attributed to a difference in communications. In the Greek city-states, the citizens (admittedly a fraction of the whole population) could come together to discuss items of mutual interest face to face. In the early days and recent history of the United States this would have been impossible due to physical distance. Today, technologies exist for having each citizen become aware of and vote on individual issues as opposed to having to depend on a representative who would generally vote as the citizen would. Tomorrow, we can readily imagine that people could not only vote on individual issues, but they would also have the wherewithal to debate individual issues. Of course, in order to have an informed public vote on complex issues, ready access to complex information in an understandable form would also be desirable.

It could be argued that being a "representative" is a full-time job and that having a full democracy would require so much time on the part of every citizen that little time would be left for other pursuits. In order to make such a democracy work, very efficient systems of information access and organization would be helpful. In addition, it must be noted that "representatives" do not actually spend all their time actually learning about and voting on issues. Much of the time, they are having their ears bent by interest groups and campaigning for reelection. If the average citizen in the United States today spent half as much time with an efficient learning system as they do watching television, they could probably become as informed as the average government representative today.

Human–Computer Interaction specialists would play a key role in making such a system feasible in the first place. In addition, such a form of "government" could also alter the way in which HCI work would take place. How might funding for grants take place in such a system? Would HCI specialists (not to mention physicists, biologists, and mathematicians) have to "sell" their specific programs of research to the general public, or would the public simply have more direct control over the relative spending levels of agencies (or pro-grams within agencies) with peer review still determining which research was funded? More radically, would intermediate funding agencies even cease to exists? Might individual citizens directly "vote" their tax dollars to specific programs or even individuals?

It seems clear that for any science (including HCI) to survive in such a radi-cally decentralized future, some new systems of checks and balances would have to evolve to replace the existing ones. Of course, one could also extend this line of thought beyond the borders of any particular country and imagine direct interpersonal negotiations operating internationally. Direct conversations among individuals of different nations, cultures, and languages would provide a rather interesting challenge for HCI professionals.

Although the aforementioned scenarios may seem seductively likely, an argument can also be made that large institutions, as self-aggrandizing entities, may act in various ways to preserve themselves so that the landscape of 2020 will not look that different institutionally from the landscape of 1995. The forces mentioned earlier for greater centralization and for greater decentraliza-tion could balance each other and provide for an institutional landscape that, in broad outline, is not that different from what we see today.

**Both Increased Centralization and Decentralization**

Finally, it could also be argued that 2020 will find not a mean shift along the centralization/decentralization continuum, but an increase in variance so that there will be both more centralization and more decentralization. Personally, I believe this to be the most likely outcome. In 2020, we will see a wider spread of institutional sizes; there will be more huge conglomerates as well as many more entrepreneurial individuals doing business globally via the Internet. This will mean that HCI specialists will be operating in a wide variety of contexts. Some will be still be in large companies and universities, whereas many others will be operating in one- or two-person consulting companies. We will also see a greater number of HCI professionals operating in the context of government. This large variance in context should facilitate the most explosive growth in HCI. Large institutions, for instance, may foster the perfection of recognition technology, whereas small ad hoc teams may foster the greatest growth rate of new ideas in the field and new (in some cases niche) applications for technolo-gies.

**Other Changes in Institutional Context**

Regardless of the type of institution in which the HCI professional finds her- or himself in 2020, there will be significant changes in the way business is carried out. Although schedules and budgets will still be a major limiting and defining role in the application of HCI, major changes in the role of HCI will also occur. Perhaps the most significant of these will be that institutions (in all sizes) will perform a much greater proportion of their decision making on the basis of cost/benefit analyses rather than just cost analyses.

Although institutions today purport to make decisions based on "the bottom line," in fact, this is the exception rather than the rule. Costs tend to be very well-defined and contained temporally, spatially, and organizationally. Benefits tend to be relatively widespread in time, space, and among subsystems. Due to today's limitations of information and computational power (as well as custom), most decisions are made on the basis of costs alone, or, at best, a tradeoff between costs and a very speculative and limited set of benefits.

Perhaps a recent example from a large corporation will be instructive. In a certain computer company, the costs of any work done to make a product more usable were carefully (and rather easily) tracked. Every dollar of additional expense for usability not only added to the cost of development, but, according to cost formulae, meant an estimated additional dollar of maintenance costs, a dollar of service costs, a dollar of manufacturing costs, and $5 of sales costs. These additional estimated costs translated to a $20 increase in estimated price. Although historical data abounded to show across products that these mean value relationships generally held (although extending that relationship to delta increments within a product was unsubstantiated), there was no historical data to indicate what the impact of the increased usability would be on revenue and other benefits such as reduced true service and sales costs.

In today's organizations there are many trends toward taking a broader view of the impact of decisions across the organization. New technology will make the tracking of benefits easier to measure and manage. This change toward cost/benefit analyses will have a significant impact on the perceived utility of HCI work. In addition, workers, consumers, and management will all be more aware of the impact of usability on sales, productivity, turnover, and morale. Many projects and products will cross traditional boundaries of industry versus university and even company versus company. Indeed, many products and projects will be carried out by ad hoc coalitions of semiautonomous individuals and small teams. Some of these members may be from universities, some from large companies, and some small independent contractors. Mechanisms will exist for distributing resulting benefits more or less equitably across organizations and individuals so that such "ad-hocracies" will work socially, financially, and legally.

The development of mechanisms to deal with "credit assignment" is a very challenging task. In a group design activity (perhaps the most important phase of development), many people are interacting simultaneously in complex ways. Recent tools may assist in keeping track of design rationale, but how should we assign relative "credit" to the team member who initiates an idea, who points

out a flaw, who fixes the flaw, whose well-timed joke relieves tension enough
for the breakthrough to occur, and whose ill-timed joke is merely disruptive?
One solution, of course, is to reward the entire team equally regardless of perceived contribution. How does HCI, as a field, then "prove" its contribution
under such a scheme? More generally, how does one measure and improve
process?

There are other issues dealing with context that may greatly impact the way
in which HCI work is carried out in the future. First, we may see a shift away
from the 20th-century insistence toward the mechanical (as Ellul, 1964, called
"technique") back toward more acceptance of the unpredictable and organic.
Thus, systems may be conceived of as optimally consisting of some combination of organic (e.g., human beings) and inorganic (e.g., computers) subsystems.
Today, the unstated ideal HCI system is often to have zero humans and all computers (thus avoiding the need for any interface whatsoever). There is substantial evidence that such "all-automatic" systems are not "better" in any reasonable sense of the word. In today's world, humans are tolerated, but only if their
operations can be predicted and controlled. By 2020, there will be a paradigm
shift so that systems will be designed under the assumption that humans (i.e.,
workers,) have knowledge, positive motivation, and creativity and this will have
a profound impact on the way in which HCI systems are designed.

A look at disasters such as the Three-Mile Island nuclear accident consistently reveals that an attempt was made to reduce the role of the human being to
procedure follower. Because there are always unanticipated consequences of
any complex system operating in a real, complex, and changing environment,
this inevitably means that people who are uninformed about the basic, underlying dynamics are suddenly thrust into the position of having to make a decision
that is not covered by the "rules" and for which they have neither situational
awareness nor an appreciation of the underlying system dynamics. Whereas
this is most dramatic when complex systems like airplanes or nuclear power
plants fail, it is no less true of the millions of more minor corporate decisions
made every day from the assembly line to the board room (cf. Garson, 1988).
By 2020, we will have abandoned the myth of the efficacy of detailed centralized planning, whether by a totalitarian government, a corporation, or a university.

## HCI Specialists

Whereas today some may think of a "human-factors expert" or an "HCI expert"
as a fairly specialized position, by 2020 we will see a whole variety of subspe-
cialists in HCI categorized along a number of dimensions based on the users,
contexts, tasks, technologies, systems, and methods. Thus, for instance, we
may see specialists in the following: USERS—the blind or the quadriplegic;
CONTEXTS—systems in police work or telephony monitoring; TASKS—

browsing or problem finding; TECHNOLOGIES—speech recognition or virtual reality; SYSTEM TYPE—Apple or IBM; METHODS—heuristic evaluation or EPIC. Given the complexity of our tasks even today, it is as though the medical field only had a category called "medical doctor." Although in 2020 there will still be a place for the "general practitioner," most will be practicing one of the many subspecialties in HCI.

Surprisingly, two new (perhaps outrageous) HCI specialties that may evolve are ACI and CCI—namely, Animal–Computer Interaction and Computer–Computer Interaction. With the growing ubiquity of computing and two-career families, we may expect that a variety of devices will attempt to monitor, control, train, and yes, even entertain, animals in their owners' absence. Such devices will only be successful to the extent that ACI expertise is brought to bear. Of course, historically it has often been true that technology better suited to the physical aspects of animals has resulted in economic benefits. For example, in the middle ages, early horse harnesses used for plowing fields were placed much too high to provide optimal leverage and, furthermore, tended to cut off blood flow through the horse's neck! Simply making a harness better suited to a horse resulted in a four- to fivefold increase in workload (White, 1962). New electronic technologies for animals will have to be designed with the widely disparate sensory, cognitive, and motoric abilities of animals as well as their physiologies in mind.

However, in what sense could "Computer–Computer Interaction" be considered a subspecialty of HCI? The answer is that today's computing systems and communication systems are so complex and have so many levels that few (if any) individuals understand all these levels and their interactions. Heterogeneous networks and systems are already described by some experts with terms from social interaction; for example, *flaky*, *finicky*, and *robust*. Useful hints on how to proceed with complex systems include at least as many heuristics as algorithms (see, e.g., Krol, 1994). In the near future, testing end-to-end application-level integrity in heterogeneous networks will look less and less like the logic testing of early simple circuits and protocols and more and more like the testing of computer–human interactions that rely on sampling, statistics, and intuitions about likely difficulties.

The increasing specialization is interlinked with several other trends described in this chapter. First, our methods will show greater refinement; that is, rather than having one parameter for a model of retrieval speed from short-term memory, we will have more differentiated models that indicate how the parameter changes as a function of age, material, physiological state, and so on. Second, HCI investigators will use an integration of intuitive, experience-based knowledge and more formal knowledge. To the extent that one becomes personally familiar with a particular set of users, contexts, tasks, and so on, the HCI expert's advice will be more apropos. Third, there will be a greater use of participative design and a greater integration of HCI expertise into the overall development process. This requires that the HCI expert devote time to client-management, trust-building, and team-building.

Only by long-term commitment to one or two projects at a time can the HCI expert have an early, broad, and profound effect on system design rather than last-minute fixing of the surface features of a system. Although various experts will specialize along a number of orthogonal dimensions, the greatest power and success will probably come from those who specialize in understanding the evolving needs of a specific set of users/contexts/tasks rather than those specializing in a specific technology or method.

### Tools to Measure Tradeoffs

Another issue dealing with context in 2020 has to do with the ability of professionals to explicitly measure and trade off among having a special tool, an individual tool, and a generic tool across users, times, and applications. More-or-less religious debates abound today in HCI about the relative value of having a "consistent" user interface across applications or the importance of keeping an interface consistent with what has already been shipped. A tool optimized to a particular application, for a particular user, task, system, and context is theoretically the most powerful. It is senseless, however, if the user will then be required to have to learn something totally different in a short time. A facility to individualize interfaces also seems desirable, yet users often learn about interfaces from their peers. This process is certainly impeded if everyone has a different interface. Although today these issues are recognized, no one has the necessary data to make an informed decision about the tradeoffs. By 2020, such issues will be largely decided by fact, not fiction. (This is not to say, of course, that there will be one answer regardless of situation.)

### Standards

A final issue that might well be included under context concerns standards. Although improved CSCW techniques will make reaching a consensus more efficient, this will be balanced by an explosion of potential technological options with the net result being that standards and guidelines will continue to lag considerably behind the state of the art in HCI. The fate of large institutions will obviously have a major impact in how the standards process operates. Although this section has dealt with the impact of context on HCI, it should also be noted that the various successes of HCI over the next few decades will also help determine which of these contexts is most likely to prevail. For instance, how easy we make it for every citizen to produce *effective* multimedia communication will play a large role in determining the extent to which such channels of communication are centralized or decentralized.

## METHODOLOGIES

No one methodology will come to "dominate" the field of HCI; instead, significant progress will be made on several fronts simultaneously.

### Heuristic Evaluation

Heuristic evaluation will be improved in many ways. First, Latent Semantic Index-like methods will be used to "assemble" optimal candidate evaluators across the world for the evaluation of a particular product. Rather than having to rely on a written description of an interface, these experts will have interface descriptions downloaded to their individual computing devices running "interpreters" where they will be able to "use" the proposed system. Improvements in computer modeling will enable the experts to experience the timing aspects of the real system fairly accurately. In addition to receiving the interface, they will also receive video-graphic descriptions of the contexts of use and the users. Various cognitive tools will aid the experts to use this information to adopt various successive viewpoints and thereby discover a greater total number of potential problems with the interface (cf. Desurvire & Thomas, 1993). Some individuals will become quite expert at doing this type of heuristic evaluation, especially within given domains—they will improve their skills by seeing what problems other experts uncover and by receiving feedback more quickly about what happens in the usability lab and in the real world. Tools will enable them to retrieve and match their predictions with actualities in the field and therefore learn effectively to modify their internal models of user behavior.

### Formal Models

Formal models will be expanded to deal much more effectively with many issues of parallelism, motivation, emotion, and error handling. In addition, we will have stable overall parameter estimates for many important model parameters; in fact, in many cases we will even have metamodels that will indicate how these parameters change as a function of the user group (e.g., age) as well as the context of use (e.g., to deal with fatigue effects). Not only will these models be more accurate; the "front ends" to these models will be greatly improved so that almost all the intellectual work required to use a formal model will be that required by the HCI work itself (as opposed to interface peculiarities). We are already beginning to see the use of dynamic visual presentations to explicate how models operate in addition to textual or mathematical descriptions. Formal models are also beginning to deal better with the emotional (Teasdale & Barnard, 1993) and parallel (Kieras et al., 1994) characteristics of human behavior. However, the formal models of 2020 will still not be able to deal completely with interpersonal behavior, problem solving, or richly multimedia I/O. On the other hand, there will be tools to aid the HCI expert's intuitions in these areas. For instance, we will be able to see the behavior of interacting agents with Virtual Reality technologies. Although this will not correctly model the impact of other (real) people, it will help the HCI expert notice the possibility of impact and help guide her or his intuitions. In addition to the refinements of formal models due to the incorporation of more accurate individualized parameter estimates and the extension to deal with motivational and emotional aspects of behavior, we would expect the trend toward more "user-friendly" front ends on systems such as SOAR, GOMS, the Human Information Processor, ACT-R, and other formal models to continue.

Perhaps the biggest change in formal models, however, will be a greater reliance on automatic methods. At this point, an analogy to recognition technology may be in order. Many early attempts at handwriting recognition and speech recognition relied heavily on human input and intuition as to the exact structure of the program. Then these programs were tested on a relatively small amount of data and in some cases, "tuned." More recent reliance on neural networks, Hidden Markov Models (HMMs) and other adaptive programming techniques that "train themselves" on massive amounts of data has resulted in much better classification and recognition results. With an increasing amount of data on human performance and the increased computer power available, we may well expect to see models developed with fewer "built-in" assumptions about how human cognition "works." SOAR might be viewed as a step in this direction. How much one wants to begin with an explicit model, of course, depends on purpose. If the primary objective is to test an explicit model of human cognition, then having many built-in assumptions is key. However, if one's primary objective is to maximize predictability or productivity, it might well be the case that fewer built-in assumptions is key, as it has been in the development of recognition technology. Ideally, the technology of Artificial Intelligence will also advance to the point where the "implicit" structures that arise in such adaptive models will become more explicable as well.

## Usability Studies

In the arena of usability studies, there has been a significant problem in the past with studying interpersonal communication systems because it involved two or more subjects, thus complicating the problems of comparability. Failing to take into account all the actors in a complex social interaction can result in gross suboptimization. In the near future we may see the use of "avatars" (cf. Fahlen, Brown, Stahl, & Carlson, 1993) or simulations of humans that can be parametrically controlled in order to provide the context of a realistic social situation. Virtual reality can also be used to simulate various aspects of the physical environment that would be otherwise difficult to portray to the human subjects.

We can also expect that increases in computation, along with increases in our basic understanding of various subject effects, will result in different experimental paradigms, not only in the field of HCI but also in experimental behavioral and medical science generally. For example, we will have better models of the space of possibilities for user interfaces within a given domain and therefore be able to test out various options more systematically. We would also expect to see a more widespread use of maximum likelihood statistics rather than classical statistics and, along with that change, sequentially varying experimental designs. In other words, as evidence is gathered that makes the superiority of one user interface option over another apparent to a cost-justifiable extent, the experimental design will change "online" so that other options are explored in more detail.

Other changes in experimental methodology will be possible due to the increased computational power available in 2020. For instance, we can also modify tasks and instructions online to keep the user focused on those features and functions of an interface about which there is some doubt. In a typical usability study today, only a small fraction of the user's behavior actually informs us about user interface design choices that are at issue. Much of the user's time is spent interacting with those aspects of the interface that we have already discovered work well or recovering from errors arising from those aspects of the interface that are already known to be problematic. By having a flexible task structure, a number of predefined interface options readily available, and a means for an expert system or a human experimenter to intervene quickly, it will be feasible for the subject in a usability study to spend a much greater proportion of his or her time behaving in an optimally informing fashion.

## Participative Design, Development, Deployment and Use

As explored by Turner, Lee, and Atwood (this volume), participative design is not enough—Users need to be involved throughout the development process. This trend will probably increase through 2020 but, in addition, new technology will enable improvements in this process. First, we are continuing to develop new techniques for participation and to test their effectiveness; second, we are learning more about the design process; third, there is a trend toward continuous quality improvement in industry and other institutions. In line with this philosophy, we would expect to see most commercial software shipped with on-board measuring devices of various kinds to help evaluate software's effectiveness and improve future versions. This process of continuous improvement will be more readily accomplished with future software because having the users specify their goals will be a relatively painless and necessary part of the use of most applications anyway. The same mechanisms can be used to give intelligent help to users when they explicitly ask for help as well as to give guidance on more efficient ways of doing things, thus helping to overcome the paradox of the active user (Carroll & Parson, 1987). Not only keystrokes but voice and video will be available to the HCI expert so that the actual contextual factors impacting use and usability can be observed. Factor analysis, multiple regression, and other techniques requiring substantial data and computing power can be applied. Walston and Felix (1977) used a similar technique to determine what actually impacted software productivity. The use of such automatic techniques obviously raises a whole set of ethical issues as well.

Interdisciplinary teams working on new software will thus have huge amounts of usability, learnability, and usefulness data from the field to help guide their redesign efforts. User companies and other institutions may begin to measure the effectiveness of their system purchases as well. Some future usability work may well model itself after epidemiology by comparing and contrasting many variables across institutions, users, contexts, and so on to get new ideas about what constitutes usable software.

The huge reductions in the cost of computing and communication that we ought to see by 2020 should also allow virtual design teams to work together despite differences in time and locality. Additionally, there may well be "framing" devices to help diverse groups of people understand each other better and work together more effectively (Thomas, 1980). One suggested scheme would continue to make common goals a salient part of the overall stimulus field while participants were disagreeing about details. Another would allow one participant to find out about another participant's viewpoint, hopefully thus providing some insight into their perspective. A primitive version of this feature exists now in America Online, which allows participants in a multiparty conversation to view profiles of the participants during the ongoing dialogue. Newer augmented reality technology could provide a much richer version of this via audio or visual supplements during the ongoing design dialogue. By understanding where other team members are coming from, there should be a decrease in needless debate and an increase in overall efficiency.

Augmented reality and virtual reality technologies could not only help design team members understand each other's context; the same technology could be used to aid the entire design team understand better contexts of use. This would be very useful for the design of complex systems that involve multiple interacting parts from different context. It would assist all the stakeholders of a system to more quickly appreciate each other's viewpoints.

## New Methods

We will also see many new usability methods by 2020. Obviously, all the methods to be invented in the next 25 years will not be foreshadowed in this chapter! However, a sampling of current trends may give some insight into what such methods may be like.

One new concept recently developed is that of "Expectation Agents" (Girgensohn, Redmiles, & Shipman, 1994). Here, the interface embodies intelligent agents that "expect" certain user behaviors at certain points in the interactions with the computer. When these expectations of user behavior are violated, the user may be queried about their intentions and a message sent back to the design team that indicates the context, what the user was trying to do, and how the designer's expectations were violated. This is just one example of how systems themselves may be instrumented to provide better usability. Another key concept here is that feedback will be provided from actual performance to the designers and keyed to particular points of the design intent.

One method of guessing (it can be little more) what new methods may be developed in HCI is to analogize from breakthroughs in other sciences. For example, a whole new level of understanding (and "applications") in physics came about from a consideration of what was at first a rather insignificant and curious effect called radiation. Is there a similar breakthrough of effectuating "mental power" that is foreshadowed now by some phenomenon that seems a mere curiosity? Could we use real-time MRI, for example, to teach people new patterns of thought or via biofeedback guide people online to much higher levels of performance?

In genetics, the ability to splice, duplicate, and reinsert genes made study orders of magnitude more efficient. Is there some analogous method that can be developed for "splicing" behavioral sequences or underlying beliefs that will allow us a much greater "signal to noise" ratio is experimental study? Interestingly, Ohnemus and Biers (1993) found that a retrospective method of think-aloud protocol led to more comments that were useful to developers than the usual concurrent think-aloud technique. Perhaps there is a way to extend this technique so that users may be put in the same "state" on multiple occasions.

## METAMETHODOLOGY

One of the most significant trends in software today is having diverse application programs and even diverse hardware systems share data. This trend will undoubtedly continue. In a similar fashion, an advance in CSCW may be the development of systems that allow human–computer systems to share information across representations and stages in the development process. In the past, one of the principal problems in software development occured because much of the information that went into design is not adequately transferred to those doing coding or maintenance. To the extent that we are successful at developing methods for quickly and efficiently communicating large bodies of information (and/or making that portion of information that is needed at a particular time easily accessible and obvious), the huge body of knowledge that is required to work in a particular field no longer becomes the limiting factor in effectiveness. Instead, one can imagine a world in which the individual's ability and desire to do certain kinds of mental functions would become the more important factor in determining work. If this becomes the case, metamethodologies may be developed which guarantee that, at any given point in the development of a complex system or in its ongoing evolution, the appropriate combination of people, information, physical resources, and methodologies are brought to bear at each point in time.

In order to handle such a system, a number of social as well as technical challenges must be met. For instance, it may be necessary to assign credit in a fairly complex fashion. Any given worker may move in and out of several projects during the course of a work year. On the other hand, if work is bid-organized so that at any given moment each person is working on the thing that is most consonant with their interests and abilities, extrinsic motivation may become relatively less important than is the case today.

## TECHNOLOGIES

The ubiquity of computing will allow usability engineers to spend much more of their time in actual contexts of use living with real users. In addition, most systems will be shipped with codes that will be used to identify usability problems in actual use. In some cases, these problems can be fixed "automatically" by changing the interface or making suggestions to the user. In more complex cases, these problems will be fed back to development/maintenance teams who will prototype new solutions, program, test, debug, and download them.

In addition to behavior on a computer system, in many cases, the contents of conversations among individuals and the dispositions of relevant paper documents can also be correlated with online behavior. Advances in machine vision and speech recognition will make this possible and in many cases cost-effective. This will provide usability experts with the data to deal with a more complete picture of the user's behavior.

Whereas computers have existed for decades, it was only when power books became cheap enough and small enough that they found their way into a wide variety of ordinary business meetings. By 2020, the changes in the cost, size, and power requirements of computing and storage along with the decreasing costs of bandwidth (and possibly more intelligent staging algorithms) will mean that computers are not only usable virtually everywhere; they may be used unnoticed everywhere. This raises the very real possibility that anytime anyone is interacting with anyone such an interaction may be supported by access to a very large, very "intelligent" computer system, and recorded for later manipulation and playback. Obviously, this can have profound social impacts and raises a whole new and thorny set of issues for the HCI professional.

## The Impact of Noncomputer Technologies

Although we would expect C/C technologies to have the most profound impact on the field of HCI in 2020, there are a number of other technologies that may have an equally widespread impact.

For example, it may well be possible to "test" people genetically for a whole variety of behavioral and cognitive predispositions. Such tests could be used to partial out effects that are now difficult to disentangle. This trend may be synergistic with the compilation of huge amounts of data on each person's experience. This data, in turn, may be retrievable in an intelligent fashion. The net result from the narrow perspective of an experimenter may be that the two largest sources of intersubject variance in behavior, namely, genetics and previous experience, may be partialed out, either statistically or by experimental design. This is especially possible if one imagines a broadband ubiquitous network. Thus, the experimenter could offer money to volunteers on a worldwide basis that met a set of highly stringent criteria for equating groups, individual by individual, as opposed to by group mean. (Can anyone seriously believe that if group A has 3 people with 10 years experience and 3 with zero years experience it is "equivalent" to a group of 6 people each of whom has 5 years experience?) Such a system could help inform a number of comparisons routinely made today (e.g., expert vs. novice) that are highly problematic to interpret. For instance, experts may generally be more intelligent, creative, or persistent, as well as having more relevant experience.

Obviously, the implications of the existence of such a widely available system of human classification according to genetics and experience go far beyond the narrow concerns of more efficient experimental design! Nor are these the only possibilities that might "benefit" the HCI scientist narrowly but have very problematic ethical and social implications. Advances in understanding the biochemical as well as the electrical nature of the brain could make it possible for people to be induced into a variety of brain-states that would impact their emotions and abilities. Although such impacts might be thought to be limited to "volunteers," history seems to offer little evidence to support such a jocund view. Employers currently offer free caffeine during work hours yet restrict (in essence) the use of nicotine and prohibit the use of alcohol. Is it such a stretch to imagine that if "safe" drugs become available that increase various kinds of productivity, employers may not encourage such drugs? Perhaps a drug will be found that simultaneously increases the ability of people to do what they are told and limits their ability or desire to question what they are told. What proportion of administrators and managers that you have known would object to the use of such a drug?

From one perspective, the free choice of drugs and/or electrical stimulation on the part of the end user may make the design of an "optimal" interface much more difficult because it may introduce additional variability. Conversely, if a system measured a person's brain state in some manner and dispensed various drugs and/or electrical stimulation in order to ensure that every user was in some "optimal" state for using a system, it might make the job of the interface designer easier. A very different perspective on human nature might hold that the individual adult user is the best judge of what, in the context of their life, is best for them at any given moment. These are social/political issues that are certainly difficult to deal with but not addressing them does not mean they will disappear. The real question is whether HCI professionals will aid in resolving these issues.

An explosion of new technologies will allow new interface possibilities. Some of these may be currently foreseen, but many more probably cannot be at this point in time. Expected technologies in widespread use would include virtual reality, interactive TV, "smart" phones, augmented reality, machine vision, automatic speech recognition, speech synthesis, person identification (via voice, behavioral patterns, location technology and machine vision), broadband communications, personal communication systems (i.e., access to the electronic matrix anytime and anywhere), agent-oriented software, and seamless paper/electronic interfaces. This means that to solve any given set of design challenges, an HCI expert will have a host of technically possible and economically feasible choices. On the one hand, ultimate systems could be much better; on the other, an HCI expert will either have to spend the largest proportion of her or his time learning about new technologies or will have to work as a part of a larger interdisciplinary team.

## Impacts of C/C Technologies

If we now examine those technological changes that are more specifically geared toward computers and communications, per se, one emerging technology that could have a profound impact is machine vision. Basically, this technology allows the computer to interpret events visually. Most primitively, it includes OCR, handprinted character recognition, and soon the recognition of handwritten text as well as fax to e-mail conversion. This will certainly aid in the seamless conversion between paper and electronic information (cf. Ishii, 1990). In addition, however, there are many other possibilities for machine vision. It can certainly aid in identifying who is where at any given time.

Further, machine vision will be capable of "seeing" things that humans cannot. Machine vision could potentially, for example, scan an office building and find the average flux of motion over time, thus giving some summary statistic about the average energy level of the people in the building. It could also be used to identify features, perhaps unstated or unstable that define a person's preference about the kinds of entertainment that they prefer. Machine vision is also the key to unencumbered gestural virtual reality systems.

Another major new technology impacting human–computer interaction will be the development of "artificial intelligence." Most of what is now called "artificial intelligence" might more accurately (if less compactly!) be called "artificial ways of emulating human intelligence at the I/O level." The attempt to enable machines to play chess, parse sentences, give expert advice, recognize speech, or transcribe handwriting to ASCII mimic things that humans can do. The economic force driving this work is that human labor costs for such tasks are high relative to what a machine might do. Although many clues for how to do such tasks have come from psychological studies of humans, the goal is to give the same or better outputs, not necessarily to simulate the underlying mechanisms that people use.

On the other hand, we can also imagine a very large family of tasks that machines might do via "artificial intelligence" that human beings not only do not do today, but that we have not even tried to do. Such tasks may be so orthogonal to the abilities springing from our biological natures that we do not even perceive them as problems to be solved. Nevertheless, just as machine vision may well be applied not only to categorizing handwritten characters as we do, but also to finding patterns humans would never perceive, so too may auditory processors, natural language processors, and other kinds of pattern discoverers be applied to huge bodies of data; the results may well be the discovery of interesting patterns. It may well be the case, for instance, that there are subtle patterns of human behavior that are critical to improved models of cognition. By *subtle* I mean patterns that are correlated in time and space in such a nonmonotonic fashion that human investigators would probably never even look for them, let alone perceive them. The implications of this possibility on our ability to predict human behavior could be anything from nothing to profound.

## SCOPE OF PROBLEMS ADDRESSED

In order to deal effectively with many of the new interface possibilities, HCI experts must expand their expertise to deal with a much broader range of human behaviors than menu selection and keyboard entry, which dominated much of the initial HCI work. Obviously, this trend is well underway! Compare, for instance, the proceedings of a recent CHI Conference with the Gaithersburg conference. In addition, communications will make computing virtually ubiquitous. The corollary of these two trends, however, may be startling to the HCI community: The expertise of HCI in 2020 will be highly relevant to almost every major enterprise of society! This will include art in the broadest sense, education, recreation, management decisions, and public policy in addition to the more obvious examples of process reengineering and the design of new products. Computers and communication technologies could be (and in most cases will be) a major part of the aforementioned activities. The "bottleneck" to meeting the full potential of new systems will be the interface between this new technology and the humans whether "performance" is measured in terms of insight, enjoyment, or productivity in the narrower sense.

It is difficult to imagine many human activities in 2020 that might not involve computers in some way (cf. Kellogg et al., 1992; Stuart & Thomas, 1991). Awakening could involve a computerized scan of favorite music tied to scanned brain state. Lower fat breakfast choices could be reinforced with appropriate positive imagery. Commuting could certainly involve multimedia and data access. Occupations, recreation, medicine—human–computer interaction will be everywhere.

Although it may seem surprising that what is viewed by some as a "narrow specialty" of the 1970s will be a major activity of human society a half-century later, we should remember that writing and telephony began their use in the hands of a small minority of individuals. The ubiquity of the impact of HCI may be reflected primarily in a larger number of experts, a more efficient access to expertise, or both. In any case, the efficacy of HCI will be a prime determinant of almost every human activity by 2020.

## CONCLUSIONS

The field of HCI will advance in major ways by 2020, but the challenge may grow even more as both the scope of relevant applications and the design space of possible solutions expand tremendously. On the one hand, we may expect new methods, new technologies, better education, and a more widespread use of reasonable development methodologies to aid the HCI expert of the future. On the other, many of these same new technologies will also mean that there will be explosive growth in the breadth of contexts that HCI will address. Furthermore, change itself is fast-paced and increasingly so. This means that although it is vital for the HCI specialist to "know" about the particular users, tasks, and contexts that they are dealing with, these very factors will be ever-changing. We may well find that 2020, for the HCI professional, exemplifies the purported ancient Chinese curse: "May you live in interesting times."

Perhaps how this balance between higher demands and better tools eventually settles will depend, ironically, on applying the principles of HCI to HCI itself. This theme was touched on in the discussion of providing more "user-friendly" front ends to formal tools. However, we may expand the concept to all the activities of HCI as a profession. As Landauer (1995) pointed out, productivity in the "service sector" only increases about 0–2% per annum with "business as usual." In those few cases where the principles of HCI are applied, however, per annum increases in productivity of 30% are reasonable. Suppose that HCI professionals themselves applied the principles of HCI to their own individual and joint work in every aspect of the profession (even if, as developers often complain when asked to apply HCI, they are too busy). An interesting exercise, left to the reader, is to calculate the possible impact on the productivity of HCI by applying HCI over the next 25 years. This is basically the difference between 1.01 to the 25th power versus 1.3 to the 25th power. (Hint: the difference is significant.)

## REFERENCES

Carroll, J., & Rosson, M. (1987). Paradox of the active user. In J. Carroll (Ed.), *Interfacing thought: Cognitive aspects of human–computer interaction.* Cambridge, MA: MIT Press.

Curtis, B., & Hefley, B. (1994). A WIMP no more: The maturing of user interface engineering. *Interactions, I.1,* 22–34.

Desurvire, H., & Thomas, J. C. (1993). Enhancing the performance of interface evaluators using non-empirical usability methods. *Proceedings of the 37th Annual Meeting of the Human Factors and Ergonomics Society* (pp. 1132–1136). Santa Monica, CA: Human Factors and Ergonomic Society.

Edwards, R. (1979). *Contested terrain.* New York: Basic Books.

Ellul, J. (1964). The technological society. New York: Random House.

Fahlen, L., Brown, C., Stahl, O., & Carlson, C. (1993). A space based model for user interaction in shared synthetic environments. *Proceedings of INTERCHI 93* (pp. 43–48). New York: ACM.

Garson, B. (1988). *The electronic sweatshop.* New York: Penguin.

Girgensohn, A. Redmiles, D., & Shipman, F. (1994). Agent-based support for communication between developers and users in software design. *Proceedings of Knowledge-based Software Engineering.*

Ishii, H. (1990). Team work station: Towards a seamless shared workspace. In *Proceedings of CSCW '90.* New York: ACM.

Landauer, T. K. (1995). *The trouble with computers: Usefulness, usability, and productivity.* Cambridge, MA: MIT Press.

Nielsen, J., & Landauer, T. K. (1993). A mathematical model of finding usability problems. *Proceedings of INTERCHI 93* (pp. 206–213). New York: ACM Press.

Nielsen, J. (1994). Enhancing the explanatory power of usability heuristics. *Proceedings of CHI '94* (pp. 152–158). New York: ACM Press.

Ohnemus, K. R., & Biers, D. W. (1993). Retrospective versus concurrent thinking-out-loud in usability testing. *Proceedings of the 37th Annual Meeting of the Human Factors and Ergonomics Society* (pp. 1127–1131). Santa Monica, CA: Human Factors and Ergonomics Society.

Teasdale, J. D., & Barnard, P. J. (1993). *Affect, cognition and change: Re-modeling depressive thought.* Hillsdale, NJ: Erlbaum.

Thomas, J. (1980). The computer as an active communications medium. *Proceedings of the 18th annual meeting of the Association for Computational Linguistics* (pp. 83–86).

Walston, C. E., & Felix, C. P. (1977). A method of programming measurement and estimation. *IBM Systems Journal, 16,* 54-73.

White, L., Jr., (1962). *Medieval technology and social change.* Oxford: Oxford University Press.

# Chapter 9
# Independent Iterative Design:
# A Method That Didn't Work*

Jakob Nielsen
*SunSoft*

Marco G.P. Bergman
*Delft, The Netherlands*

## INTRODUCTION

This is the story of a failed project. Actually, the project was not a complete failure because we did learn several important lessons and also succeeded in arriving at a fairly good user interface design for our system. However, we wasted considerable resources and failed to collect important data. This chapter tells the story of what we tried to do and why it didn't work. If nothing else, these lessons can help other people avoid making the same mistakes. Even better, this project may serve as a datapoint to build up an understanding of what can cause usability projects to fail.

### Case Study Domain

For the case study reported here, we developed a user interface to a personal communications service (PCS) running on a home computer with a graphical user interface and a mouse. The PCS concept involves giving a customer a single, individualized telephone number that is independent of the physical loca-

*The project described in this chapter was done while the authors were at Bellcore. The authors would like to thank Sheila Borack for substantial help in scheduling subjects and other matters related to the running of the experiment. We also thank the six Bellcore user interface designers who spent significant time redesigning our interface and the seven usability specialists from Bellcore, NYNEX, and U S WEST who provided estimates of potential usability goals for the project. Because these people were in some sense the real subjects for our study, ethical considerations prevent us from listing their names, but they know who they are: thank you! The authors thank Tom Landauer for helpful comments on earlier versions of this manuscript.

tions in which that customer wants to receive calls. The customer contacts the PCS service to let it know to which physical location it should forward calls at any given time. In our experimental version, a person can set up time-dependent calling profiles, having calls forwarded to the office during working hours, to a car phone during commuting hours, and to a home phone during the rest of the day, with a different calling profile in force on weekends. Further features involve the use of screening lists to ensure, for example, that only calls originating from a few specified telephone numbers will be allowed through late at night or that collect calls from certain numbers will be accepted automatically. In addition to the currently active profile, our experimental system also supports a master profile that contains parameters that should always be in force (e.g., *never* allow calls from a certain telephone number). PCS systems did not exist at the time of our project, so we could not compare our experimental interfaces with competitive designs.

## Design Method

To develop our experimental PCS user interface, we used a variant of iterative design called *independent iterative design*. In iterative design, several versions of a user interface are developed, one after the other, with each version being subjected to user testing (or other evaluation) to find usability problems which are supposed to be fixed in the next version. Often, the first iteration is a previous version of the product but as mentioned, we did not have any existing products to work from. Independent iterative design follows the basic iterative design project model but involves a new designer for each new version in the hope of utilizing the new designer's ideas to achieve more rapid interface improvements than one might expect from the traditional method where a single designer is being asked to modify his or her previous design.

Our hope was that independent iterative design would bring some of the benefits from parallel design to the iterative design method (Nielsen, 1994; Nielsen et al., 1993). In parallel design, several designers are asked to design an initial version of the interface, working concurrently and independently of each other. As shown in Figure 1, both iterative and parallel design are based on a "version 0" conceptual idea describing what the interface is supposed to do in general terms. When using parallel design, this initial vision is instantiated in multiple alternative ways using some kind of rapid design approach, meaning that version 2 (the first unified interface design) can be based on the best ideas from each of the designers. Parallel design is thus a method for speeding up time to market by exploring multiple design alternatives simultaneously.

A potential weakness of parallel design is the "waste" of resources inherent in having several designers work at the same time, even though not all of their design ideas will be used (or even subjected to user testing if the first version to be tested is the unified design in version 2). An alternative that would allow for

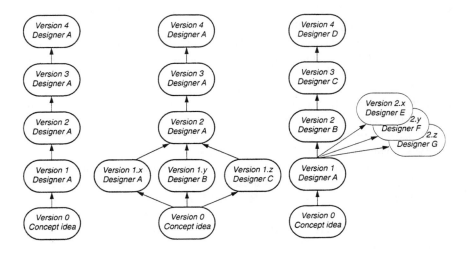

Figure 1. Models of standard iterative design (left), parallel design (middle), and the process used in this case study (right).

the involvement of multiple designers and yet not waste effort is the independent iterative design method shown to the right of Figure 1.

Independent iterative design follows the standard approach of iterative design in only considering one interface design at a time, subjecting each version to user testing, and designing the next version based on the previous version and a list of the observations from the user test. Independent iterative design differs from standard iterative design by involving a new designer for each version, in the hope that this new designer will be emotionally independent of any attachment to the poorer aspects of the previous design and thus more willing to change it to accommodate the users. For the purpose of our research project, we had four different designers redesign version 1, resulting in version 2 (the one used as a basis for version 3) as well as three additional versions (marked version 2.x, 2.y, and 2.z in Figure 1) that were not used in the actual project but were used to get research data showing the spread in usability one might expect from a single iteration, depending on who redesigned the interface. Thus, versions 2.x, 2.y, and 2.z are only shown in Figure 1 to give an overview of the structure of our entire research project: for the purposes of development projects, we envision independent iterative design to encompass only the versions marked 1, 2, 3, etc. in the figure. Specifically, our design method involved showing each new designer the previous versions according to our project model as well as a list of the usability problems found by our user tests. Thus, referring to Figure 1, Designer A only had access to the conceptual vision of the system; Designer B (and E, F, and G) had access to version 1 and the list of problems in version 1; Designer C had access to versions 1 and 2, as well as the list of problems in these interfaces; and, finally, Designer D had access to versions 1, 2, and 3 and their usability problems. The design designated as version 2 and used as the basis for version 3 is simply the first one from our pool of four people doing redesigns of version 1.

A variant of parallel design is called *diversified parallel design* and involves having each of the designers in the parallel design stage optimize their interface for one specific platform or user group without consideration of the other platforms or user groups that need to be supported eventually. For example, one designer can design an interface explicitly for novice users while another designs an interface explicitly for expert users, thus allowing both designers to explore various aspects of the design space more fully than would be possible with an all-encompassing design. Of course, the eventual unified design will have to support all intended users and platforms, but it will be able to do so based on a refined set of ideas. For our current study of independent iterative design, we did not attempt to include any principles from diversified parallel design, as we did not expect any benefits from changing the intended users or platforms several times throughout an iterative design process. One possibility for at least partially supporting the goals of diversified parallel design in iterative design would be to select one focus for the initial several versions (e.g., to get the interface optimized for novice users using a small screen). After the interface has been polished with respect to this initial goal, one could then add features or design aspects in later versions to support secondary users or platforms while testing to make sure that usability was not significantly impaired for the primary users.

## TEST METHOD

For each iteration, the interface was tested with 10 subjects, except for the first version, which was tested with 24 subjects to ensure higher reliability of the baseline measure and the last version which was tested with 11 subjects due to overbooking of subjects. Including pilot subjects and subjects for whom no data is available for various reasons such as system crashes,[1] we tested a total of 99 subjects. The subjects were undergraduate students in fields other than comput-

---

[1]For the purposes of this particular research project, we wanted "clean" performance times, so we discarded the data from subjects who encountered programming errors in our interface implementations. For most usability engineering purposes, much can still be learned from user tests where the system crashes.

er science who were paid $24 to participate in a test session lasting about two hours. For each test, the subjects were first given a general introduction to the personal computer and allowed to practice using the mouse. They were then given 11 test tasks one at a time and asked to perform them as quickly as possible.

When subjects felt that they had completed a task, they pressed a button marked "task is finished." For each task, the computer recorded the time from when the subject was given the task statement to the time he or she pressed the button. After the subject had pressed the button, the experimenter informed the subject whether the task had indeed been performed correctly. If the user had not performed the task correctly (e.g., forwarding calls to the home phone during the wrong time of day), the experimenter would simply say that an error had been made (but not *what* the error was), and ask the subject to try again. If the subject performed the same task incorrectly multiple times, the experimenter would provide progressively more helpful hints until the subject had performed the task correctly. The "task time" referred to later is the time from the task statement was first presented to the final, correct completion of the task. The "error count" is the number of times the user pressed "task is finished" without in fact having finished the task correctly. Finally, after the last task, the users was given a simple subjective satisfaction questionnaire.

In addition to monitoring whether subjects performed the tasks correctly, the experimenter noted any difficulties they had while using the interface. After the complete set of tests with a given interface, the experimenter wrote a report listing the usability problems observed in the interface, and this report was given to the designer in charge of designing the next version. Of course, there is never any guarantee that all the usability problems that could have been found with an infinite number of test users were in fact found during our tests. To get an estimate of what proportion of the total set of usability problems we actually found, we fitted a mathematical model of the finding of usability problems to the data from the first study with 24 subjects (Nielsen & Landauer, 1993). The model parameter 1 (which indicates the proportion of usability problems found with a single subject) was .24, indicating that 94% of the usability problems would be found by testing the first 10 subjects. Thus, even though we did not test all our interfaces exhaustively, we did test them sufficiently to find the overwhelming majority of usability problems.

## DEVIATIONS FROM RECOMMENDED PRACTICE

There are several ways in which our method in this project deviated from what we would normally recommend for a traditional development project without a research component. As discussed further later, the independent iterative design method itself turned out not to be as successful as we had expected, so we would probably not recommend it for future use. Furthermore, as mentioned earlier, the "extra" versions 2.x, 2.y, and 2.z (shown in Figure 1) were only designed for research purposes to study the spread in possible redesigns and were deliberately kept secret from the designers of later versions, and one would normally not develop these extra versions just for the fun of it.

Each version of the system was tested with at least 10 test users even though we would normally recommend using no more than 3 to 5 users per version. For the sake of our research project, we needed many test users in order to get reasonably tight confidence intervals on our usability measures, but with respect to learning about the interface in order to design the next version, not much additional information was gained after the first 5 or so users.

Finally, we kept the test tasks constant for all iterations in order to collect comparable data from all the tests. If we had not needed to measure the same tasks for research purposes, we would definitely have eliminated some features from the system in later versions because they proved to be too difficult for the users. Specifically, the distinction between different levels of calling profiles (the global profile that was always active, and the current profile that potentially modified it for the duration of some specific situation) turned out to be too complex for most users. Normally, we would have simplified the system in later versions by eliminating features that turned out to complicate the interface more than they were worth. However, since one of our research goals was to compare measured usability across versions, we needed to keep the same set of test tasks for all versions, and it was thus necessary for all versions to support the same tasks. This approach corresponds to the case where one is developing a system to match a prespecified set of unchangeable requirements but it not recommended in general, because it is preferable to change the systems' feature set as more information is gathered about the users' needs.

## RESULTS

Table 1 shows the results of user testing of all the interface versions developed in this study. Results are given both as absolute numbers (the actual mean values measured for each version) and in relative terms, showing how much better or worse each version was compared with version 1. Basically, the table shows that the interface did not get very much better from each iteration to the next, because usability remained fairly constant. Also, the four different redesigns of version 1 (versions 2, 2.x, 2.y, and 2.z) were about equal in usability, ranging from 90% to 106% of the usability of version 1 (with a standard deviation of only 7%), meaning that the best redesign was only 17% better than the worst. Because all designers did about equally badly, the poor performance of the redesigns was not the fault of any individual designer but was due to an inherent problem.

| Version | Task Time in Minutes | Relative Performance | Error Count | Relative Error Rate | Subjective Rating, 1–7 | Relative Subjective | Overall Usability |
|---------|---------------------|---------------------|-------------|---------------------|------------------------|---------------------|-------------------|
| 1 | 46 (12) | 100% | 7.9 (3.4) | 100% | 3.0 (0.7) | 100% | 100% |
| 2 | 46 (12) | 99% | 6.8 (3.3) | 116% | 2.9 (0.8) | 104% | 106% |
| 2.x | 52 (19) | 88% | 9.8 (4.4) | 80% | 2.9 (1.1) | 105% | 90% |
| 2.y | 50 (12) | 92% | 8.9 (2.9) | 88% | 2.9 (0.7) | 106% | 94% |
| 2.z | 48 (12) | 95% | 8.3 (2.9) | 95% | 2.6 (0.6) | 115% | 101% |
| 3 | 49 (15) | 93% | 7.7 (3.2) | 102% | 2.7 (0.5) | 111% | 102% |
| 4 | 38 (11) | 120% | 6.3 (3.1) | 125% | 2.8 (0.5) | 110% | 118% |

Table 1.  Measured usability of the various versions of the PCS interface. The columns for relative usability show how much better (or worse) each version was compared with version 1, with usability values of more than 100% indicating improvements in usability, and values smaller than 100% indicating reduced usability. Subjective ratings were on a scale from 1 to 7, with 1 indicating the best rating. The transformation of the users' subjective satisfaction ratings to an approximate ratio scale was done using the method described in Nielsen and Levy (1994). Overall usability was computed as the geometric mean of the relative usability measures of performance, error rates, and subjective satisfaction. Numbers in parentheses indicate standard deviations. Versions 2.x–2.z were not used in the design of versions 3-4.

The 18% improvement in overall usability from version 1 to 4 (through three iterations) corresponds to an average improvement of only 4.9% per iteration. Even though the improvement from version 1 to version 4 is small, it is statistically significant at the $p < .05$ level for overall usability and for task time according to two-tailed t-tests.

Nielsen (1993) found median improvement in measured usability of 38% per iteration across four usability engineering projects. Bailey (1993) studied eight designers who designed an average of four versions (three iterations) of an interface, achieving a mean improvement from the first to the last version of 47% in needed interventions by the experimenter and 35% in task time. Thus, overall usability was improved by 41% over three iterations, corresponding to 12% per iteration. There are several possible reasons for Bailey's comparatively modest improvement in usability per iteration: Only three subjects were used per iteration, the designers did not watch the users directly but only via a video recording (and the subjects were not asked to think aloud, making it hard to infer usability problems from the video), and not all designers were experienced in usability.

Compared with improvements of 38% in four studies and 12% in a study where the designers had suboptimal conditions, the 5% improvement in overall usability per iteration in our study is appallingly low. We expect that a major reason for the slight improvements was the use of new, independent designers for each iteration. The designers were not satisfied with simply fixing the usability problems in the previous design; each designer also introduced fundamentally new design ideas to the design, causing a phenomenon we call *design thrashing*, where the design never stabilized or reached a polished state. Each of the many new ideas may have had its positive aspects, but each time the design changed direction, new usability problems were introduced, causing the overall design to be no better than its predecessor with respect to measured usability.

Although measured usability improved very little over the iterations, we do believe that the conceptual structure of the interface improved considerably, making the final version a much better basis for further development than the initial version. The test users seemed to perform the tasks in the initial version without ever getting a real conceptual understanding of the structure of the system. In contrast, users of the final version exhibited a reasonable degree of understanding, and would therefore presumably be in a better position to improve their performance further with time and expand their usage of the system to additional tasks. However, we do not have any firm evidence for this effect because we did not measure the users' conceptual understanding of the system.

In retrospect, not collecting data on the users' conceptual understanding of the system was a major mistake. Such data could have been collected by a simple method such as a multiple-choice test, or through more advanced methods for the elicitation of mental models. The main reason we did not collect these data was that we were focused on primary usability attributes like performance, errors, and subjective satisfaction. These primary attributes must be the main goals of usability engineering, because one does not really care what model the users have of the system as long as they like it and are efficient using it. Thus, measures of the users' conceptual model are only secondary usability attributes of a system. However, secondary measures may be very important for assessing the potential for further improvement in usability and in choosing between alternatives for further development. In our case, version 4 was better than version 1 on all three of the primary usability attributes we measured, so there is no doubt that we will recommend version 4 as the basis for further development since we also believe that it is better structured conceptually.

As a thought experiment, assume that the only two versions we had available were version 1 and version 2.y. Based on our test user observations, we believe that version 2.y was conceptually better than version 1 as the foundation for further development, so we would probably have recommended version 2.y over version 1, even though it was worse according to measured usability. One of the main slogans of discount usability engineering is "any data is data," and of course, our informal observations of the test users is some form of data that can be used to justify such a recommendation. Even so, we would have been more comfortable if we had had quantitative data to back up our impressions of the users. Therefore, we would recommend collecting data on the users' conceptual understanding of the system in cases where the underlying system is complicated.

## SETTING USABILITY GOALS

It is often recommended to base usability engineering efforts on measurable usability goals that are specified before the interface is designed (Whiteside, Bennett, & Holtzblatt, 1988). Such goals will allow project management to assess whether the product has reached a sufficient level of usability or whether additional iterations will be required before release. Typically, four levels of usability are defined: current usability of existing or competing products, minimum acceptable level, planned (or target) level, and theoretically optimum level.

For our case study, no current or competing systems were in service, and we could thus not estimate a 'current level' of usability for the tasks supported by our system. Given the lack of data from other systems, we found it very difficult to establish usability goals for the other three levels of usability. We finally settled on a method based on averaging usability estimates from a group of usability specialists. To get these estimates, we first defined eight concrete tasks that could be performed with the system. These same tasks were later used in user testing (along with a few additional tasks used to achieve closure in the test scenario), making it possible to compare the usability goal levels with measured usability for the various versions of the user interface.

A list of these tasks was then sent to a number of usability specialists who were asked to provide their best estimate as to how long time it would take users to perform each of the tasks. The usability specialists provided their estimates independently of each other. For each task, the usability specialists estimated three performance times: the longest time that the task would be allowed to take for the interface to be acceptable, the task time they thought we should plan to achieve, and the fastest they thought it would be possible to perform the task with a "perfect" user interface. The usability specialists had to provide these three time estimates purely on the basis of a functional description of the tasks (i.e., what the user was supposed to achieve by doing the task) and were not given access to any specific user interface designs to support the tasks.

We gathered usability estimates from seven usability specialists with an average of 11 years of usability experience (ranging from 5 to 20 years) who indicated that measurable usability goals were only specified on 21% of the projects they had been involved with. The 21% was a mean value, with individual replies ranging from 0 to 90%. Given the large variation in the replies, this estimate of projects using measurable usability goals corresponds reasonably well with the estimate of 38% for the same number in an earlier survey of a different group of usability specialists (Nielsen, 1992). Both surveys thus indicate that most projects currently do not employ measurable usability goals even though a sizable minority of projects do.

For the eight tasks described to the usability specialists in our survey, a system with minimally acceptable usability was estimated as requiring a total task time of 42 minutes (mean across the seven estimates). Target usability was estimated as 21 minutes, and the theoretically optimal usability was estimated as corresponding to a task time of 12 minutes. In fact, with our initial design (version 1) users required 39 minutes to perform the eight tasks, whereas they could perform them in 31 minutes with our final design (version 4). Thus, even though we did not achieve our target level of usability, we did produce an interface that could in theory be shipped. In reality, though, we were only developing an experimental interface.

As might be expected, the usability goal estimates varied widely. Averaged over all estimates, the standard deviation of individual task time estimates was 61% of the mean estimates. The variability was slightly smaller for estimates of the overall time to perform all eight tasks, with a standard deviation of 51% of the mean estimate.

For each individual usability specialist there was a very high correlation between his or her estimates of the three levels of usability. Averaged across the usability specialists, the correlation between minimally acceptable task times and target times was .90, the correlation between target times and theoretically optimal times was .93, and the correlation between minimally acceptable times and theoretically optimal times was .79. In other words, the specialists assumed certain tasks to be inherently difficult and take longer than certain other, inherently easy, tasks, no matter whether a minimally acceptable, target, or optimal user interface was to be used to perform the tasks.

In spite of considerable disagreement regarding the absolute goal levels for the various tasks, the usability specialists exhibited a fair degree of unanimity regarding the ratios between the three levels. Averaged over all pairs of estimates, the ratio between the minimally acceptable time and the target time was 2.1 (with a standard deviation of 0.5). The ratio between the target time and the theoretically optimal time was also 2.1 (but with a higher standard deviation of 1.0). Further research will be necessary to determine whether the ratio of 2.1 is in fact reasonable, or whether it was just used by our usability specialists for convenience.

Looking at the degree of agreement between the individual usability specialists, the average correlation between estimates was .28 for minimally acceptable usability, .44 for target usability, and .41 for theoretically optimal usability. These results indicate that the usability specialists tended to agree much more with respect to the target and optimal usability levels than with respect to the minimally acceptable usability level. One possible reason for this finding may be that target usability and theoretically optimal usability depend only on the task descriptions and the possible user interfaces envisioned by the usability specialists. The minimally acceptable usability depends on these two factors as well but also on the usability specialists' assessment of market conditions. Possibly, usability specialists disagree more about what the market will bear than about what constitutes a good user interface for a task.

Further analyzing the degree of agreement between the usability specialists, Kendall's coefficient of concordance, $W$, was .39, .52, and .50 for minimally acceptable, target, and theoretically optimal usability, respectively. All three values are statistically significant at the $p < .01$ level ($X^2 = 18.9, 25.3, 24.3$; $df = 7$), indicating that the agreement between the usability specialists is not just due to chance.

Thus, even though there is considerable disagreement between the usability specialists as to the absolute time it might take users to perform various tasks, there is reasonable agreement as to what tasks are difficult and what are easy. This observation suggests a possible alternative approach to estimating usability goals: One could include some tasks for which user performance was known on the list of tasks for which the usability specialists were asked to provide goal estimates. Usability goals for the new tasks could then be set relative to known values for good and acceptable usability for the old tasks, using conversion factors given by the estimated values for the old and new tasks. Similarly, assuming that one of the three goal levels can be empirically determined for a given task, the other two might be derived using the 2.1 ratios found to describe our specialists' estimates. We have not yet tried doing so.

## COST ACCOUNTING

The time and money spent on this project were as follows:

- *Subjects*: 99 subjects at $24 = $2,376.
- *Research assistants*: Scheduling subjects: 2 hours per subject = 198 hours. Running user tests and summarizing data: 4 hours per subject = 396 hours. Assuming that the loaded cost of a research assistant is $30 per hour, the cost of research assistants related to the subjects was $17,820. Learning about the PCS problem domain: 80 hours. Writing up reports from user tests for use by the designers: 20 hours per design = 120 hours. Scheduling designers: 2 hours per designer = 12 hours. Taking notes from design sessions in order to be able to implement the redesigns: 6 hours per design session = 36 hours. Implementing the original version: 200 hours. Implementing the six redesigns: 80 hours per design = 480 hours. The cost of research assistants related to the designs was thus $27,840.

*Usability professionals*: Designing initial study and modifying it based on pilot tests: 40 hours. Supervising additional testing and reporting: 4 hours per iteration = 24 hours. Redesigning interface: 6 designers spending 6 hours each = 36 hours. Collecting usability goal estimates: 7 usability specialists at 2 hours each = 14 hours. Analyzing goal estimates: 4 hours. Assuming that the loaded cost of a usability specialist is $125 per hour, the cost of the usability professionals were $14,750.

## FIXED AND VARIABLE COSTS

The fixed costs of this project included learning about the domain, designing the initial study (including pilot testing of the tasks), gathering and analyzing usability goals, and implementing the initial design: a total of $16,466. About $8,400 of these fixed cost cannot really be considered as usability engineering costs as they would have been needed even for a development project without any usability activities (learning about the domain and implementing a solution). Thus, the additional fixed cost of iterative design as opposed to hit-and-run design were $8,066.

The variable costs per iteration included the time spent on a redesign, implementation of a redesign, testing the design with 1 pilot user, 10 regular test users, and an average of one test that could not be used due to crashes or the like (for a total of 12 test users), and the writing of the test report: a total of $6,938 per iteration.

The total cost of the project was $62,786, which included 99 test users (85 of whom provided actual data) and seven fully implemented user interfaces. One way of looking at the cost data is that each user test that provided data cost $739 (the variable cost of a single user was only $204, but then not all users provided data and there were large costs associated with planning the project and implementing the interfaces). Alternatively, one could say that each interface version cost $8,969 to implement and test on the average, when distributing the fixed costs over all the versions.

## DISCOUNT USABILITY ENGINEERING ALTERNATIVES

The project could have been made cheaper in many ways. The most obvious savings would come from eliminating versions 2.x, 2.y, and 2.z which were only included in the project for research purposes. Avoiding these three versions saves about $20,814. Further major savings would have been realized from using a prototyping tool to implement the user interfaces instead of a general-purpose, object-oriented GUI programming system. We used the general-purpose programming system to be able to implement the wide variety of design suggestions we received from our designers (it seemed that every designer came up with at least one new interface widget that was not in any standard package), but it is often not necessary (or even desirable) to support such deviations from interface standards. Based on experience from other projects using such prototyping tools, we estimate that the implementation time for both the original version and the subsequent versions could have been cut by at least 50%, corresponding to a further saving of $6,600. The data from user testing could have been analyzed much more rapidly by using a simplified approach that only considered the experimenter's notes and impressions from the tests (and not test logs as used by us), for an estimated saving of about 1 hour of experimenter time per subject, or $2,970. Finally, it would have been sufficient to run five test subjects (plus pilot subjects and 'wasted' subjects) for each version, corresponding to a further saving of $5,742.

Deducting all these savings lead to a "discount" project with a total cost of $26,660, with fixed costs of $13,346 and variable costs of $4,508 per iteration. A "deep discount" project could have been achieved by eliminating the measurable usability goals: not only would there have been no need to collect the goal estimates, but the initial design of the study would also have been easier. We estimate that these two changes would reduce the fixed costs to $9,846.

## BENEFIT ESTIMATION

As usual, benefits from usability engineering are hard to determine. One approach is to consider the value of the time saved by the potential future users. Because we are making a guess about the future, we will be conservative and assume that the user base for the system will correspond to no more than 10% of the current number of U.S. subscribers to cellular service, or about 1.3 million users. Furthermore, we will assume that the users will save the 9 minutes we measured in our test tasks during early use of the system but that they will not save any additional time during extended use. Finally, we will assess the value of the users' time as low as $5 per hour because many users will be using the system in their spare time. These assumptions translate into $975,000 as the total value of the time saved by the users.

Of course, time saved by the users does not provide direct earnings for the service providers. There is currently no data available to estimate how usability impacts sales but to get a rough idea, we will assume that improved usability might increase sales by one tenth of the increase in usability. Thus, the 13% improvement in usability in our project will be assumed to increase sales by 1.3%, or about 17,000 users. If users were to pay as little as $5 per month for the service, 17,000 extra users would mean about $1M additional revenue per year. Much of this revenue would be profit since these extra users would be marginal, meaning that the main user base would presumably have paid for the fixed costs of building and operating the service.

These two approaches estimate the benefits as being about 10–15 times larger than the costs. This benefit–cost ratio is low compared to those seen in many discount usability engineering projects, but the ratio is certainly large enough to conclude that even failed usability projects are well worth their cost.

## CONCLUSIONS

Independent iterative design cannot be recommended as a project model for usability engineering. Constantly involving new designers may lead to design thrashing where the design never stabilizes and new usability problems are introduced as rapidly as the old problems are designed away.

We estimate that the variable cost of each additional iteration in iterative design is between $4,500 and $7,000 for a medium-sized project like the one studied here. Even the 5% improvement in usability per iteration in our project may be enough to justify such a cost for many applications, and the costs should certainly be justified if our mistakes are avoided and the more typical, larger improvements in usability are realized.

## REFERENCES

Bailey, G. (1993). Iterative methodology and designer training in human–computer interface design. In *Proceedings of ACM INTERCHI'93* (pp. 198–205). New York: ACM Press.

Nielsen, J. (1992). The usability engineering life cycle. *IEEE Computer, 25*(3), 12–22.

Nielsen, J. (1993). Iterative design of user interfaces. *IEEE Computer, 26*(11), 32–41.

Nielsen, J. (1994). *Usability engineering.* Boston, MA: Academic Press.

Nielsen, J., & Landauer, T. K. (1993). A mathematical model of the finding of usability problems. In *Proceedings of ACM INTERCHI'93* (pp. 206–213). New York: ACM Press.

Nielsen, J., & Levy, J. (1994). Measuring usability—Preference vs. performance. *Communications of the ACM, 37*(4), 66–75.

Nielsen, J., Desurvire, H., Kerr, R., Rosenberg, D., Salomon, G., Molich, R., & Stewart, T. (1993). Comparative design review: An exercise in parallel design. In *Proceedings of ACM INTERCHI'93* (pp. 414–417).

Whiteside, J., Bennett, J., & Holtzblatt, K. (1988). Usability engineering: Our experience and evolution. In Helander, M. (Ed.), *Handbook of human–computer interaction* (pp. 791–817). Amsterdam: North-Holland.

# Author Index

# Subject Index